AF898576

AN ASSESSMENT OF **POTENTIAL HEALTH EFFECTS** FROM **EXPOSURE** TO **PAVE PAWS LOW-LEVEL PHASED-ARRAY RADIOFREQUENCY ENERGY**

Committee to Assess Potential Health Effects from Exposures to PAVE PAWS
Low-Level Phased-Array Radiofrequency Energy

Board on Radiation Effects Research

Division on Earth and Life Studies

NATIONAL RESEARCH COUNCIL
OF THE NATIONAL ACADEMIES

THE NATIONAL ACADEMIES PRESS
Washington, D.C.
www.nap.edu

THE NATIONAL ACADEMIES PRESS 500 Fifth Street, N.W. Washington, DC 20001

NOTICE: The project that is the subject of this report was approved by the Governing Board of the National Research Council, whose members are drawn from the councils of the National Academy of Sciences, the National Academy of Engineering, and the Institute of Medicine. The members of the committee responsible for the report were chosen for their special competences and with regard for appropriate balance.

This study was supported by contract F05604-01-C-9000, between the National Academy of Sciences and the U.S. Air Force. Any opinions, findings, conclusions, or recommendations expressed in this publication are those of the author(s) and do not necessarily reflect the views of the organizations or agencies that provided support for the project.

International Standard Book Number 0-309-09309-0

Additional copies of this report are available from the National Academies Press, 500 Fifth Street, N.W., Lockbox 285, Washington, DC 20055; (800) 624-6242 or (202) 334-3313 (in the Washington metropolitan area); Internet, http://www.nap.edu.

Copyright 2005 by the National Academy of Sciences. All rights reserved.

Printed in the United States of America.

THE NATIONAL ACADEMIES

Advisers to the Nation on Science, Engineering, and Medicine

The **National Academy of Sciences** is a private, nonprofit, self-perpetuating society of distinguished scholars engaged in scientific and engineering research, dedicated to the furtherance of science and technology and to their use for the general welfare. Upon the authority of the charter granted to it by the Congress in 1863, the Academy has a mandate that requires it to advise the federal government on scientific and technical matters. Dr. Bruce M. Alberts is president of the National Academy of Sciences.

The **National Academy of Engineering** was established in 1964, under the charter of the National Academy of Sciences, as a parallel organization of outstanding engineers. It is autonomous in its administration and in the selection of its members, sharing with the National Academy of Sciences the responsibility for advising the federal government. The National Academy of Engineering also sponsors engineering programs aimed at meeting national needs, encourages education and research, and recognizes the superior achievements of engineers. Dr. Wm. A. Wulf is president of the National Academy of Engineering.

The **Institute of Medicine** was established in 1970 by the National Academy of Sciences to secure the services of eminent members of appropriate professions in the examination of policy matters pertaining to the health of the public. The Institute acts under the responsibility given to the National Academy of Sciences by its congressional charter to be an adviser to the federal government and, upon its own initiative, to identify issues of medical care, research, and education. Dr. Harvey V. Fineberg is president of the Institute of Medicine.

The **National Research Council** was organized by the National Academy of Sciences in 1916 to associate the broad community of science and technology with the Academy's purposes of furthering knowledge and advising the federal government. Functioning in accordance with general policies determined by the Academy, the Council has become the principal operating agency of both the National Academy of Sciences and the National Academy of Engineering in providing services to the government, the public, and the scientific and engineering communities. The Council is administered jointly by both Academies and the Institute of Medicine. Dr. Bruce M. Alberts and Dr. Wm. A. Wulf are chair and vice chair, respectively, of the National Research Council.

www.national-academies.org

COMMITTEE TO ASSESS POTENTIAL HEALTH EFFECTS FROM EXPOSURES TO PAVE PAWS LOW-LEVEL PHASED-ARRAY RADIOFREQUENCY ENERGY

FRANK S. BARNES *(Chairman)*, University of Colorado, Boulder, CO
ROBERT C. HANSEN *(Vice-Chairman)*, R. C. Hansen, Inc., Tarzana, CA
LARRY E. ANDERSON, Battelle, Pacific Northwest National Laboratories, Richland, WA
GRAHAM A. COLDITZ, Harvard Medical School, Boston, MA
FRANCESCA DOMINICI, Johns Hopkins University, Baltimore, MD
KENNETH J. McLEOD, State University of New York, Binghamton, NY
KEITH D. PAULSEN, Dartmouth College, Hanover, NH
LESLIE L. ROBISON, University of Minnesota Cancer Center, Minneapolis, MN
SUSAN L. SANTOS, University of Medicine and Dentistry of New Jersey School of Public Health and East Orange New Jersey Veterans Administration Medical Center War Related Illness and Injury Study Center, Medford, MA
JAN A. J. STOLWIJK, Yale University School of Medicine (emeritus), Beltsville, MD
GAYLE E. WOLOSCHAK, Northwestern University, Chicago, IL

Consultant

DAVID R. CHALLONER, University of Florida, Gainesville, FL

National Research Council Staff

RICK JOSTES, Study Director
EVAN B. DOUPLE, BRER Director
COURTNEY GIBBS, Program Assistant
DORIS E. TAYLOR, Staff Assistant

BOARD ON RADIATION EFFECTS RESEARCH

S. JAMES ADELSTEIN (*Chairman*), Harvard Medical School, Boston, MA
HAROLD L. BECK, Department of Energy Environmental Laboratory (Retired), New York, NY
JOEL S. BEDFORD, Colorado State University, Fort Collins, CO
JAMES E. CLEAVER, University of California San Francisco Cancer Center, San Francisco, CA
SARAH C. DARBY, University of Oxford, Oxford, United Kingdom
SHARON L. DUNWOODY, University of Wisconsin, Madison, WI
C. CLIFTON LING, Memorial Sloan-Kettering Cancer Center, New York, NY
DANIEL KREWSKI, University of Ottawa, Ottawa, Canada
THEODORE L. PHILLIPS, University of California, San Francisco, CA
ANDREW M. SESSLER, Lawrence Berkeley National Laboratory, Berkeley, CA
JOHN C. VILLFORTH, Food and Drug Law Institute (Retired), Derwood, MD
PAUL L. ZEIMER, Purdue University, West Lafayette, IN

EDWARD R. EPP (Board Liaison to the Committee), Massachusetts General Hospital (Retired), Boston, MA

National Research Council Staff

EVAN B. DOUPLE, Director, Board on Radiation Effects Research
ISAF AL-NABULSI, Senior Program Officer
RICK JOSTES, Senior Program Officer
CATHERINE S. BERKLEY, Administrative Associate
DORIS E. TAYLOR, Staff Assistant
COURTNEY GIBBS, Program Assistant

Preface

In a January 11, 2001, letter from Senator Edward M. Kennedy to the Secretary of the Air Force, F. Whitten Peters, Senator Kennedy asked that the Air Force fund an independent study through the National Research Council of the National Academies "to examine the health effects of the PAVE PAWS system." Kennedy further requested that "this follow-on study (to the previous 1979 National Research Council report) should address, at a minimum, the effects, if any, of the PAVE PAWS radar over the past two decades and should also examine the validity of using continuous-wave and pulsed non-ionizing radiation biological-effects data as surrogates for phased-array non-ionizing radiation biological effects data." The offices of Senators Kennedy and Kerry, and Congressman Delehunt, participated in discussions with the Air Force and the National Research Council to establish the task that is addressed by this committee in this report.

A committee composed of individuals with engineering, biology, epidemiology, risk communication, and biostatistics expertise was established by the National Research Council to address the task. The committee heard from interested citizens in a public forum held in Sandwich, MA on May 28, 2002. In additional open sessions of the committee, the committee gathered information from the Air Force, the Massachusetts Department of Public Health, the PAVE PAWS Public Health Steering Group, various experts from universities and other institutions, and the Cape Cod public.

The committee also evaluated information provided to the committee by interested parties, surveyed the scientific literature, and conducted a preliminary statistical correlation analysis to evaluate the potential for biological and health effects from the PAVE PAWS radar. A letter report was published providing advice regarding the Air Force waveform measurement effort and a second interim letter report was published that commented on the adequacy, at that time, of available information and outlined the general characteristics of information that the committee deemed useful in its evaluation of the potential biological and health effects of the PAVE PAWS radar.

This report is an update of a prior 1979 National Research Council report (*Analysis of the Exposure Levels and Potential Biologic Effects of the PAVE PAWS Radar System*) and addresses the following:

1. The applicability of, and the level of uncertainty associated with, using data derived from cell, animal, and epidemiological studies employing pulsed and continuous-wave exposure for evaluation of potential adverse health effects following phased-array exposures;
2. The extent of the exposure of the public to electromagnetic energy from the PAVE PAWS system;
3. Potential biological and health effects of the PAVE PAWS Radar System; and
4. Recommendations for appropriate follow-on study design issues, including the strengths and limitations of the approaches suggested and the potential value of the proposed work.

Frank S. Barnes, Ph.D.
Chair

Reviewers

This report has been reviewed in draft form by persons chosen for their diverse perspectives and technical expertise in accordance with procedures approved by the National Research Council's Report Review Committee. The purposes of this review are to provide candid and critical comments that will assist the institution in making the published report as sound as possible and to ensure that the report meets institutional standards of objectivity, evidence, and responsiveness to the study charge. The review comments and draft manuscript remain confidential to protect the integrity of the deliberative process. We wish to thank the following for their participation in the review of this report:

Thomas F. Budinger, University of California, Berkeley, CA
David G. Hoel, Medical University of South Carolina, Charleston, SC
Daniel Krewski, University of Ottawa, Ottawa, Canada
Steven C. Lewis, University of Texas, Dallas, TX
Frank Prato, St. Joseph's Health Centre, London, Canada
Andrew M. Sessler, E.O. Lawrence Berkeley National Laboratory, Berkeley, CA
Zenon Sienkiewicz, National Radiological Protection Board, Chilton, United Kingdom
Bernard Veyret, University of Bordeaux, Pessac Cedex, France

Although the reviewers listed above have provided many constructive comments and suggestions, they were not asked to endorse the conclusions or recommendations, nor did they see the final draft of the report before its release. The review of this report was overseen by **John F. Ahearne**, Sigma Xi and Duke

University, Research Triangle Park, NC, and **Richard B. Setlow**, Brookhaven National Laboratory, Upton, NY. Appointed by the National Research Council, they were responsible for making certain that an independent examination of this report was carried out in accordance with institutional procedures and that all review comments were carefully considered. Responsibility for the final content of this report rests entirely with the authoring committee and the National Research Council.

Acknowledgments

The committee would like to thank Dr. Richard Albanese for directing the committee's attention to aspects of the PAVE PAWS waveform that merited investigation. Richard and Sharon Judge, Charles Kleecamp, Ron Cronin, Jim Tomlin, and Victor Vyssotsky provided many insightful comments to the committee. We thank the Massachusetts Department of Public Health, the PAVE PAWS Public Health Steering Group, the Coalition to Operate PAVE PAWS Safely, and the U.S. Air Force for providing information to the committee.

We would also like to thank the following speakers at committee information-gathering sessions: Stephen Cleary, Stetson Hall, Robert S. Knorr, Joseph Roti-Roti, Kurt Oughstun, Thomas Roberts, Daniel Wartenberg, Robert Torres, and Donald McLemore.

The committee is especially indebted to the conscientious support and guidance provided by the study director, Rick Jostes. He sought and provided important information from a number of sources and he diligently kept the committee focused on its timeline and its charge. Dr. Jostes was well assisted in the administration of the committee's work by Courtney Gibbs and Doris Taylor. Courtney Slack, a Christine Mirzayan Science & Technology Policy Graduate Fellow, provided additional valuable assistance to NRC staff.

Contents

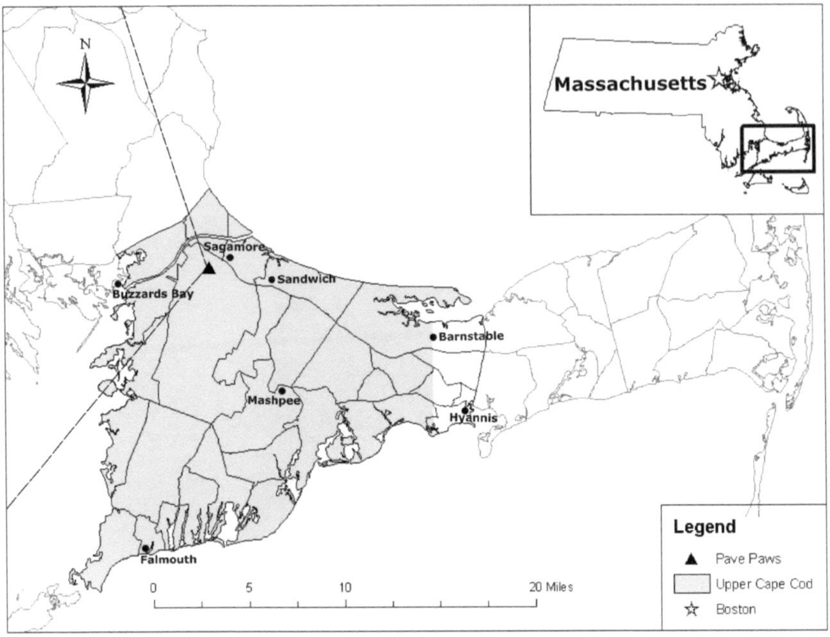

Map of Cape Cod in Massachusetts showing the "Upper Cape" as a shaded area. The Upper Cape is the portion of the Cape which is closest to the mainland and is commonly taken to extend to Barnstable. The location of the PAVE PAWS radar is indicated by the solid triangle near the town of Sagamore. The dashed lines extending from the radar indicate the approximate boundaries of the main beam when the radar is scanning (radar beam projects angled upward and to the East). From http://www.wordiq.com/definition/Cape_Cod on 10/27/2004.

Public Summary

INTRODUCTION

This report examines the potential biological and human-health effects from exposure to PAVE PAWS low-level phased-array radiofrequency energy. The PAVE PAWS radar system, part of the U.S. Air Force Space Command, is located at the Cape Cod Air Force Station in Cape Cod, Massachusetts (see map that precedes the Public Summary). The facility has been in continuous operation since 1979. "PAVE" is an Air Force program name and "PAWS" stands for Phased Array Warning System. The primary purpose of the facility is to detect and track sea-launched and intercontinental ballistic missiles. The system's secondary function is to track earth satellites and identify other space objects.

Even before the facility began operation, there had been concerns expressed by at least some members of the public regarding its safety and whether or not the facility had the potential to cause adverse health impacts. In 1979, concerns voiced by the public included the possibility of thermal effects, disruption of implanted medical devices (such as pacemakers), and secondary radiation effects from improperly grounded structures exposed to the radar. Those concerns in part led to a 1979 National Research Council (NRC) report on exposure levels and potential biological effects of the PAVE PAWS radar. That committee found that ". . . the PAVE PAWS radar may be anticipated to expose a limited number of members of the general public intermittently to low intensities of pulse-modulated microwave fields with maximal intensities of 100 $\mu W/cm^2$ or less and time-averaged intensities lower by two orders of magnitude. There are no known irreversible effects of such exposure on either morbidity and mortality in humans or other species." That committee also recommended that the Air Force conduct additional research and surveillance to evaluate the potential exposure effects of PAVE PAWS. Specifically, the 1979 report recommended:

- "Additional research is recommended to clarify further the possible effects of long-term exposure to microwave radiation at low power densities," and
- "In view of the known sensitivity of the mammalian central nervous system to electromagnetic fields, especially those modulated at brainwave frequencies, the possibility cannot be ruled out that exposure to PAVE PAWS radiation may have some effects on exposed people. Because these effects are still hypothetical, it is not feasible to access their health implications. Such assessment will require additional research and surveillance and must be addressed in future evaluations of the potential exposure effects of PAVE PAWS and other high-power-output radar systems."

The present NRC committee found no evidence that the Air Force or others followed up substantially on the above two recommendations.

Public concerns over the possibility for adverse effects from the PAVE PAWS facility have continued over the years since the time of the 1979 NRC report. In recent years, public concerns have shifted away from thermal effects of the radiofrequency (RF). Concerns instead have focused on:

- the possible biological relevance of the waveform itself;
- the inherent time delay of the phased-array radar including the secondary beams or sidelobes that are below the main beam; and
- the possible implications for health arising from the propagation of the RF energy in tissue. Some members of the public have questioned whether radiation from the PAVE PAWS system is unique such that existing safety measures may not adequately protect the public. Others have stated that the system—in spite of its unique configuration—is not that dissimilar from the other sources of RF energy to which the public is constantly exposed such as FM radio stations, TV stations, or continuous-wave radar systems.

In an effort to try to address the questions that have been raised regarding the safety and uniqueness of the system, in January 2001, Senator Edward M. Kennedy asked the U.S. Air Force to fund an independent study through the National Research Council of the National Academies "to examine the health effects of the PAVE PAWS system" and to address, in a follow-on report to the 1979 NRC report, the effects if any, of the PAVE PAWS radar over its two-plus decades of operation.

What the Committee Did

The committee undertook an extensive data- and information-gathering effort. That effort included 4 sessions that members of the public were invited to attend and at which researchers whose work was referenced as important by members of the public, or considered important by the committee, were invited to provide the committee with information. An additional meeting of the committee was held as a public forum in which interested members of the public were encouraged to present their viewpoints to the committee. In addition, there were several members of the public, who, on a number of occasions, requested that the committee review specific information they wished to be made available to the committee. Over 200 submissions of information were made to the committee by interested parties. Because there have been no studies of a phased-array system similar to PAVE PAWS in the public domain, we reviewed all the relevant available data (i.e., peer reviewed and scientifically available) in the radiofrequency range most applicable to the PAVE PAWS system (see appendix A). Further, in response to concerns raised by some members of the public that classified data might exist showing effects of a phased-array radar, a number of committee members with sufficient scientific expertise and security clearances also examined and assessed whether there was any classified research done by the U.S. Air Force that might show any evidence of biological effects with potential relevance to human-health effects of radiation similar in characteristics to PAVE PAWS.

The committee found no evidence of any classified, phased-array experiments that either were relevant to the PAVE PAWS exposure conditions or indicated a potential for PAVE PAWS human-health effects. Thus, we do not believe there is any classified data showing potential harm from the PAVE PAWS system.

RESULTS/CONCLUSIONS

The committee's conclusions address three primary areas: the implications of the PAVE PAWS waveform, the potential for biological effects, and the potential for human-health effects.

The PAVE PAWS Waveform

The recently collected waveform-characterization data that the committee reviewed has answered many questions. Based on that review and some additional statistical analyses we performed, we reached the following conclusions:

1. The PAVE PAWS narrow-band radiation is in fact similar to that of continuous narrow-band reflectors or so-called dish antennas. Those large parabolic reflector (dish) antennas are widely used for satellite earth terminals and for

radars. Both reflectors and phased arrays have time delays, and comparable size reflector antennas also have comparable delays.

2. The large number of PAVE PAWS active elements (1792) and their irregular spacing make the discrete beam formation almost indistinguishable from a continuous formation.

3. The existence and possible biological significance of precursors (additions to a signal waveform that may occur before, during, or after the signal waveform) forming would be extremely small and probably not measurable for the narrow-band PAVE PAWS system.

Potential for Biological Effects

The committee concluded:

4. Relevant data exist from experiments with animals and cells exposed under certain RF conditions that contribute to an understanding of RF biological effects and to an understanding of the potential for human-health effects from PAVE PAWS.

5. There is no risk of cancer, reproductive or developmental effects, or neurobehaviorial effects based on a comprehensive review of animal studies or studies in other biological systems. A few statistically significant biological changes have been reported from RF exposures, but the relevance of those biological changes is not known and may or may not have any impact on human health.

Potential Public-Health Effects

The committee recognizes the concerns of some of the members of the public regarding the ongoing operation of the facility, especially in light of the increase in colorectal, breast, prostate, and lung cancers that have been reported in the upper Cape over time. To date, those observed elevated cancer-incidence rates among residents of upper Cape Cod have not been adequately explained through previous investigations exploring a variety of environmental factors including PAVE PAWS. The inability of investigators to explore the possibility of health effects from the PAVE PAWS radar was due principally to the lack of PAVE PAWS RF power-density information at that time.

To determine the potential for health effects, it is important to have an estimate of exposure. One of the consistent problems in most epidemiologic studies is the lack of adequate exposure data. This was true in the relevant epidemiologic studies evaluated by the committee for other populations exposed to either pulsed or continuous radiofrequency energy. Unfortunately, there are too many limitations in those epidemiological studies to rely on them for making a determination of the potential impact of radar exposure on human health. With regard to PAVE

PAWS exposures, the historic lack of waveform characterization data and exposure data (in the form of power-density measurements) at locations where exposure to the Cape Cod population occurs has made assessment of the potential for health effects difficult. Recent waveform and power-density models and measurements by the Air Force and Broadcast Signal Laboratories have enabled some analyses by this committee and enabled a forthcoming health study by the International Epidemiology Institute.

The committee concluded:

6. The available power-density measurements are generally consistent and show that the spatial distribution of the PAVE PAWS radiofrequency energy, and thus potential for exposure, is strongly influenced by site-specific local topography and intervening terrain at any given location. The measured data show that average power densities are consistently below 0.1 mW/cm^2 and generally in the 0.001-0.01 mW/cm^2 range at locations where the public would be expected to be exposed. Measured peak levels are generally less than 1 mW/cm^2, although values as high as 15 mW/cm^2 have been found at a few elevated locations near the radar where exposure might occur. The levels of exposure can be compared to Environmental Protection Agency studies of FM and TV broadcast bands (54-900 MHz) in the 1970s. Those studies estimated that the median exposure in urban areas was 0.005 mW/cm^2 and that 95% of the urban population was exposed to less than 0.1mW/cm^2 from FM and TV broadcasts. Recent studies on cell-phone base stations in Great Britain, Canada, and Australia show RF frequencies in the vicinity of base stations ranging from 0.01 mW/cm^2 to a high reading of 2.6 mW/cm^2.

7. The potential for an individual's exposure over time is determined by how long he or she resides at any possible point(s) where exposure might occur, and the level of exposure at that particular point, which will vary with time and other factors. In spite of recent site-specific measurements and estimates of the PAVE PAWS waveforms and power densities that now exist for a number of geographic locations, there are still no data currently available to determine an individual's personal exposure to RF radiation from the PAVE PAWS radar.

8. Using information on population density, topography, and direction of the PAVE PAWS radar beam, we estimated that, based on the 1990 census, 12,773 of the total resident population (11.8% including children) of upper Cape Cod were living in the line of sight[1] of the PAVE PAWS antenna and most likely receiving some exposure from the sidelobes of the PAVE PAWS radar beam (but not the primary beam, which is angled upward). Based on 2000 census data, the

[1]Line of sight means that there are no hills between the resident and the radar that would block the radar emissions. The main beam is aimed above the population and residents in the line of sight are exposed to the sidelobes of the main beam.

estimated number of population living in the area exposed to the PAVE PAWS radar-beam sidelobes was 16,403 (12.4%).

9. Using power-density information from models provided by Mitre and recent power-density measurements and models provided by Broadcast Signal Laboratories, this committee also did its own statistical analysis. Based on our own statistical analyses, we did not identify any increase in cancer risk with exposure to the PAVE PAWS beam using peak and average power-density estimates. The analyses looked at the reported occurrences of all cancers combined on the upper Cape as well as specific cancers, including colorectal, breast (female), prostate, and lung. We are also aware of the epidemiologic investigation that is currently being conducted by the International Epidemiology Institute, but data from that study were not available to review as of the writing of this report.

10. Socioeconomic status does not appear to influence results. We performed additional analyses to see whether some indicators of socioeconomic status might influence the results (an adjustment routinely made in health or epidemiologic studies). We found that adjusting for the proportion of the population below the poverty level did not influence the results.

11. As another overall measure of health for the upper Cape Cod towns, the committee looked at premature mortality before age 75 as a useful indicator. Based on 2001 data, Barnstable, Falmouth, Mashpee, and Sandwich have lower mortality than the Massachusetts state average, while Bourne has elevated mortality.

12. Further analysis by the committee indicates that increasing duration of exposure to the PAVE PAWS radiofrequency energy has not resulted in increased incidence of cancer over time. The committee compared the standard cancer-incidence rates, or SIRs, for 5 categories consisting of total cancers, breast, colon, lung, and prostate cancer for the period of 1987-1994 versus 1995-1999 (which are the periods that the State of Massachusetts reports data) for the 5 towns in upper Cape Cod and found that there was no consistent pattern of increase. During those two time periods, a decrease in SIR was observed in 15 out of 25 SIRs, an increase in 6 out of 25 SIRs, and no change in 4 out of 25 SIRs. Again, the results indicate that increasing exposure to PAVE PAWS over time has not resulted in an increased incidence of cancer.

Summary

In summary, based on the available scientific evidence, the committee concludes there are no adverse health effects to the population resulting from continuing or long-term exposure to the PAVE PAWS radiation. In particular, the committee concludes that there is no increase in total cancers or cancers of the prostate, breast, lung, or colon due to exposure to the PAVE PAWS radiation. Further, there are many studies and data that support the finding of no health or biological effects from RF exposures. There are a number of possible mechanisms and pathways by which electric and magnetic fields could lead to changes at higher power-density levels than the public is exposed to from the PAVE PAWS radar. However, at this time, the committee has not found evidence of a mechanism shown to change biologic processes at power levels that are associated with the PAVE PAWS radar. The recent waveform-characterization data collected for the PAVE PAWS radar have also shown that they are similar to exposures from "dish" radars to which the public is also continuously exposed.

It is extremely difficult, if not impossible, to prove ultimate safety. In the United States, various forms of safety or risk assessment are used along with regulatory guidelines to ensure that facilities, products, technologies, and other factors will not pose undo risk or harm to the public or environment. The scientific community, including medical professionals, is often reluctant to call something "safe" and so often speaks of having or not having some degree of evidence of harm or lack thereof. There is also growing interest in what is referred to as the "precautionary principle," which seeks to avoid taking actions that might have the potential for harm unless a relative degree of safety can be assured. Those decisions are policy or management decisions and not solely a matter of science. This committee has focused on the scientific evidence and carefully evaluated all the scientific evidence available to determine whether there is a reasonable degree of certainty regarding the presence or absence of harm from exposure to the PAVE PAWS phased-array radar. To those who live in the vicinity of that system, no less would be acceptable.

Recommendations Regarding Further Studies

The committee was also tasked to recommend further studies if warranted. The committee recognizes that while biological responses do not necessarily translate into human-health effects, studies on the biological effects of RF exposures should be done that build upon several existing studies demonstrating a statistically significant response to RF exposure, such as the effect of radars on studies of tree growth. Future studies should approximate the PAVE PAWS exposure characteristics as closely as possible. Specifically, we recommend that studies of tree growth in the vicinity of the PAVE PAWS facility should be done. A study of long-term exposures under conditions similar to human exposures might provide useful information as to any possible mechanisms for a biological response that currently does not exist. In addition, we recommend that a replication of a central nervous system endocrine function study be undertaken to confirm or refute previous Air Force-sponsored studies showing a significant and extended influence on brain dopamine levels during low-level RF exposures similar to that of PAVE PAWS.

The Toler and other studies demonstrating a significant and long-lasting effect on serum dopamine levels does point to a biological effect that might result in a detrimental health effect. TheToler study is one of the few studies we are aware of that utilized 435 MHZ, and effects on brain activity were a major concern of the 1979 NRC review committee, so this study holds additional importance. Moreover, the study utilized a 1 KHz modulation that would not be expected to have as profound an effect as a modulation frequency similar to that of PAVE PAWS, which is in the 10-100 Hz range. For those reasons, it is recommended that this study be refined and repeated.

Finally, because of the limitations and uncertainties that exist in estimated exposure at the individual level and the number of health outcomes of interest, future health investigations or epidemiologic studies should look at exposures at both the census-tract[2] and census-block levels, and try to better estimate personal exposure and consider the types of factors known to complicate human-health investigations. Future or ongoing health studies should also specifically address possible early age of exposure and/or early age at onset of an adverse health effect. Finally, future epidemiologic studies should not be conducted unless they are expected to have sufficient statistical ability, or so-called power, to be able to detect any possible health effects in the Cape Cod population.

---

[2]For census reasons, states are divided into counties, which are in turn are divided into census tracts, which are further subdivided into census blocks. Most census tracts have between 1500 and 8000 people and they average about 4000 inhabitants. Census blocks are subdivisions of a census tract and are the smallest area for which decennial census data are available to the public.

Executive Summary

INTRODUCTION

This report, prepared by the National Academies Committee to Assess Potential Health Effects from Exposures to PAVE PAWS Low-Level Phased-Array Radiofrequency Energy is the fifth Academies report on the PAVE PAWS radar located at the Massachusetts Military Reservation in Cape Cod, Massachusetts. Two previous reports evaluated the engineering and potential for population exposure associated with the PAVE PAWS radar in 1979, just prior to the radar becoming operational.[1] Two recent letter reports by this committee provided advice on the Phase IV waveform measurement effort and evaluated the status of information available to the committee.[2]

Individuals in the Cape Cod community have raised concerns that exposure to PAVE PAWS radiation represents a unique type of radiation exposure due to the possibility that there are:

[1]NRC (National Research Council). 1979a. Analysis of the Exposure Levels and Potential Biologic Effects of the PAVE PAWS Radar System. Washington, DC: National Academy Press; NRC. 1979b. Radiation Intensity of the PAVE PAWS Radar System, Engineering Panel on the PAVE PAWS Radar System, Final Report. Washington, DC: National Academy Press.

[2]NRC. 2002. Letter Report to the Department of the Air Force from the Committee to Assess Potential Health Effects from Exposure to PAVE PAWS Low-Level Phased-Array Radiofrequency Energy: Recommendations for Phase IV Measurements. Washington, DC; NRC. 2002. Interim Letter Report to the Department of the Air Force from the Committee to Assess Potential Health Effects from Exposure to PAVE PAWS Low-Level Phased-Array Radiofrequency Energy: Adequacy of Available Research Data (see Appendix A of this report). Washington, DC.

- Particularly "steep" rise-times associated with the PAVE PAWS waveform, and
- Overlapping wave fronts originating from multiple antennas on the PAVE PAWS radar.

It has been suggested that if those temporal characteristics exist, then the PAVE PAWS radar signal could give rise to precursor formation that might result in greater exposure to biological tissue.

TASK FOR THE REPORT

This committee has been tasked to provide an update of the 1979 Academy report that includes a discussion of:

1. The applicability of, and the level of uncertainty associated with, using data derived from cell, animal, and epidemiological studies employing continuous-wave exposure for evaluation of potential adverse health effects following phased-array exposures;
2. The extent of the exposure of the public to electromagnetic energy from the PAVE PAWS system;
3. Potential biological and health effects of the PAVE PAWS radar system, and
4. Recommendations for appropriate follow-on study-design issues, including the strengths and limitations of the approaches suggested and the potential value of the proposed work.

The Executive Summary addresses the task items in order of appearance above.

1. The applicability of, and the level of uncertainty associated with, using data derived from cell, animal, and epidemiological studies employing continuous-wave exposure for evaluation of potential adverse health effects following phased-array exposures.

APPLICABILITY OF USING NON-PHASED-ARRAY DATA

Because there were essentially no PAVE PAWS data provided to or discovered by the committee that specifically addressed PAVE PAWS health effects, the first task of the committee was to determine whether continuous and pulsed radiofrequency (RF) energy research are adequate for determining the biological and potential health effects of the PAVE PAWS phased-array system. As stated in our interim report, the committee concluded that certain waveforms and power-density levels are applicable in the evaluation of biological effects and their relevance to potential health effects. For the purposes of this study the committee has reviewed cell, plant, animal, and epidemiological data derived from pulse-modulation RF exposures. A major focus of the committee was a review of data with exposure characteristics defined in the committee's interim report (for a description of these exposure characteristics, see the committee's interim report in Appendix A).

The conclusions above were based in part on the committee's understanding of the PAVE PAWS radar characteristics as noted below.

RADAR CHARACTERISTICS

While minor differences have been measured in sidelobe energy characteristics, phased-array radar characteristics are substantially equivalent to those of dish radars. That observation is based on the following results of measurements made during the Phase IV waveform measurement investigation carried out by the Air Force Research Laboratory:

- PAVE PAWS is a narrow-band system (5 MHz) consistent with its theory of operation.
- Phased-array antennas and reflector antennas have measurable time delays at wide angles. Both arrays and dishes of comparable size have comparable delays.
- The large number of PAVE PAWS active elements (1792) and their irregular spacing make the discrete beam formation almost indistinguishable from a continuous formation and thus, the role of individual elements is usually not evident.
- Precursors are additions to a signal waveform, and may occur before, during, or after the signal waveform. They have been calculated and measured, but only for wideband signals—typically, 10,000 MHz bandwidth in dispersive media. They may decay slowly, but only after significant attenuation in the cellular media. For the narrow-band PAVE PAWS radar, any precursors would be extremely small and probably not measurable. Precursor formation is directly related to bandwidth (rise time), and dispersion, but not to electric field slope (V/m/nsec).

UNCERTAINTIES INVOLVED IN USING NON-PHASED-ARRAY DATA FOR EVALUATION OF POTENTIAL ADVERSE HEALTH EFFECTS FOLLOWING PHASED-ARRAY EXPOSURES.

There are a number of uncertainties associated with using data derived from cell, animal, and epidemiological studies employing continuous-wave and pulsed RF exposures for evaluation of potential adverse health effects in humans following phased-array exposures. Those uncertainties relate to the significance of the waveforms, their repetition rates and power levels on the biological systems, and the significance of induced biological responses as a predictor of human-health effects. Those uncertainties are summarized at the end of Chapter 1.

2. The extent of the exposure of the public to electromagnetic energy from the PAVE PAWS system.

Various measurements of PAVE PAWS power-density levels in the Cape Cod community have been made by the Air Force and by citizen groups. PAVE PAWS power-densities were also modeled by Mitre Corporation, a contractor to the Air Force. A recent analysis by Broadcast Signal Labs included measured and modeled power-density information. After evaluating the information, the committee concludes that:

- The measured data show that average power densities are consistently below 0.1 $\mu W/cm^2$ and generally in the 0.001-0.01 $\mu W/cm^2$ range at locations where the public would be expected to be exposed.
- Measured peak levels are generally less than 1 $\mu W/cm^2$, although values as high as 15 $\mu W/cm^2$ have been found at elevated locations near the radar.
- Power-density measurements recorded by different groups at different times within the communities surrounding the Cape Cod PAVE PAWS radar are generally consistent with each other and with modeled results.
- The measurements show distribution patterns that are strongly influenced by site-specific local topography and intervening terrain at any given location.
- The modeled data are generally consistent with the measured data with the exception that the modeling results are higher overall than the measured results. However, this is to be expected from the approximations that were made in setting up the simulations that do not include scattering and other losses.
- The available measurement data and models of the PAVE PAWS power-density emissions provide a good first-order characterization of the spatial distribution of the exposures occurring throughout the communities of Cape Cod.
- In spite of the measurements and estimates of the PAVE PAWS waveforms and power densities, currently there are no data for estimating personal exposure at the level of an individual.
- Using information on population density, topography, and direction of the PAVE PAWS radar beam, the committee estimated that the proportion of the population of upper Cape Cod whose primary residence is in direct line of sight[3] to the PAVE PAWS beam was 11.8% in 1990 and 12.4% in 2000.

[3]Line of sight means that there are no hills between the resident and the radar that would block the radar emissions. The main beam is aimed above the population and residents in the line of sight are exposed to the sidelobes of the main beam.

3. Potential biological and health effects of the PAVE PAWS radar system.

After a review of the information available to this committee at this time, the committee concludes that there is no health hazard from exposure to the PAVE PAWS radiation. There are however, a few statistically significant biological responses that have been reported. Since biological effects do not necessarily translate into adverse health effects, the implications of the biological observations are not known. At this time it is not known which characteristics of RF radiation, at the power levels associated with the exposures to the general public from the PAVE PAWS radar, are biologically significant, if any.

The committee's conclusions from its review of biology and epidemiology relevant to the PAVE PAWS radar are outlined below.

BIOLOGY

Mechanisms

There are a number of physical mechanisms by which electric and magnetic fields of radiofrequencies are known to lead to biological changes. There are no known physical mechanisms that predict a physical interaction with tissue resulting in a biological response at power densities, on the order of 1 $\mu W/cm^2$, that are associated with the PAVE PAWS radar.

Animal Studies

- Evidence from animal studies does not provide a case for an increase in cancer risk with low-level RF exposure.
- There is no evidence of adverse effects on mammalian reproduction and development during exposure of animals to RF at levels that do not produce a significant temperature rise in tissues.
- There is some evidence of low-level RF exposure being capable of significantly reducing dopamine (a neurotransmitter in the brain involved in the control of movement) levels. Given the magnitude of the reduction observed, and the extended duration of the effect, we believe these studies merit replication.
- Alteration of behaviors, both learned and unlearned, can occur at exposure levels as low as 1 mW/cm^2 (a thousandfold higher than PAVE PAWS exposures); however, those changes appear to be reversible with no long-term or permanent effects.

Other Biological Studies

- Few biological studies have been completed using RF exposures whose physical characteristics are similar to the PAVE PAWS system; however, a wide variety of biological studies have been undertaken based on RF exposures in the frequency range of PAVE PAWS. No biological experiments have employed the exposure durations that some Cape Cod residents have experienced.
- No endpoints traditionally associated with carcinogenesis have been found to be reproducibly affected by RF exposure in cell systems.
- Long-term exposure (multiple-year) studies on plant growth undertaken with radar exposures very similar to that of PAVE PAWS showed a dose- and distance-dependent decrease in tree growth, but there is currently no means for extrapolating from the biological responses reported in these studies to any health effect in humans.

EPIDEMIOLOGY

Epidemiology of Radiofrequency Exposures Other than PAVE PAWS

- Studies reported in the peer-reviewed scientific literature describing health outcomes among human populations exposed to pulsed or continuous-wave radiofrequency exposures vary in scientific quality and provide conflicting results.
- The methodological issues associated with those investigations primarily relate to the lack of precision in the exposure assessments, which seriously limits the use of the data in making a determination of the potential impact of radar exposure on human health.

Epidemiology of the Cape Cod Population

General Cancer Incidence Rates on Cape Cod

- Using the upper Cape Cod cancer-incidence review of 1986-1994 from the Massachusetts Department of Public Health, statistically significant increases in the standardized incidence ratio (SIR) are seen on Cape Cod for the following cancers: colorectal (SIR 112), breast (SIR 110), prostate (SIR 130), and lung (SIR 112). The observed elevated cancer-incidence rates among residents of upper Cape Cod have not been adequately explained through subsequent investigations.
- As an overall measure of health for the upper Cape Cod towns, premature mortality before age 75 is a useful indicator. Based on 2001 data, Barnstable, Falmouth, Mashpee, and Sandwich have lower mortality than the Massachusetts state average, while Bourne has elevated mortality.

- Contrasting the calculated SIRs for 5 categories (total cancers, breast, colon, lung, and prostate cancer) during 1987-1994 versus 1995-1999 for 5 towns in upper Cape Cod demonstrated no consistent pattern of increase. During those two time periods, a decrease in SIR was observed in 15 out of 25 SIRs, no change in 4 out of 25 SIRs, and an increase in 6 out of 25 SIRs, suggesting that increasing duration of the presence of the PAVE PAWS radar has not resulted in increased incidence of cancer. Analyses covering longer durations may be required to confirm this initial observation.

Committee Analysis of Cancer Incidence and the PAVE PAWS Radar

- Health-outcome data on cancers in the population surrounding the PAVE PAWS radar have been published in the upper Cape Cod cancer-incidence review of 1986-1994. A statistical analysis was conducted by this committee using the 1986-1994 data and peak and average power density as the measure of residential exposure. The committee evaluated the occurrence of all cancers combined, as well as specific cancers, including colorectal, breast (female), prostate, and lung, and did not identify any increase in cancer risk with exposure to the PAVE PAWS radar.
- Adjustment for socioeconomic status (utilizing the percent of population below the poverty level) had little impact on the observed lack of correlation between the radar and health outcomes evaluated.
- The population size of upper Cape Cod places limits on the ability to conduct well-designed and well-conducted research to evaluate health outcomes associated with PAVE PAWS; however, moderately low levels of increased risk ranging between 40% and 100% (i.e., relative risks between 1.4 and 2.0) could be investigated, with acceptable statistical power, for some of the more common forms of cancer.

4. Recommendations for appropriate follow-on study-design issues, including the strengths and limitations of the approaches suggested and the potential value of the proposed work.

RECOMMENDATIONS

- Because of the limitations of human-exposure assessment, confounders, and range of health outcomes, the committee does not recommend that further PAVE PAWS health investigations be conducted unless they integrate: (1) exposure assessment and health outcomes at the census-block level; (2) strategies to quantify personal exposure in addition to residential location; and (3) appropriate considerations of potential confounders. In addition, it is recommended that future geographical correlation studies should be carried out for age-specific strata that address early age of exposure and/or early age at onset of adverse health outcomes.

- To determine the levels of extremely low-frequency (ELF) sideband energy in the PAVE PAWS radar signal, a Fourier analysis should be made of a representative search pulse pattern, including the dead (maintenance) interval. Non-linearities in tissue impedance (which need to be determined) might convert the sideband energy to ELF energy that may have biological effects. This relatively simple analysis could identify or rule out potential effects from ELF that may be generated by the PAVE PAWS RF exposure. Many other radar systems employ diagnostic and calibration pulse sequences and it should be noted that many radars may have small ELF sideband energy; the possible ELF is due to the waveform characteristics, not to the phased-array antenna.
- Theoretical work is needed to establish connections between the physics of the electromagnetic fields and changes in chemical reaction rates and the binding of molecules to membranes. Such investigations would lend insight into possible mechanisms of biological effects of RF exposures. The strength of such theoretical investigations is that they can help determine which of a large number of possible experiments are likely to identify changes that lead to biological effects and how such effects are a function of parameters such as the field strengths, frequency, and exposure times. The limitation of such theoretical work is that the models that are analyzed may not accurately describe the problems of interest and may miss some important parameters.
- Selected studies on the biological effects of RF exposures are warranted, particularly those that build upon existing studies demonstrating statistically significant responses to RF exposure. While biological responses do not necessarily translate into health effects, it is important to follow up on known biological effects to determine if they relate to an identifiable health effect.

Therefore, we suggest two types of biological studies:

1. Studies are recommended that should include large-scale genomic and proteomic screening to identify gene and protein expression patterns in cells and animals after exposure to simulated PAVE PAWS radiation, both at levels approximating peak exposure levels to the Cape Cod population and at higher power levels to identify potential threshold power densities for biological effects. The committee recognizes that the limitation to these *in vitro* and animal studies is that they may have limited applicability to the human situation. However, the strength of these studies is that they will provide information covering a wide range of cellular activity in a limited number of experiments and the information gained may be useful in generating mechanism-related hypotheses of the effects of PAVE PAWS exposures in cell and animal systems.
2. Studies of plant growth in the vicinity of the PAVE PAWS facility, such as tree ring width before and after the facility became operational, are also recom-

mended. While these studies have the limitation of not being directly applicable to human health, they have the specific strength of involving long-term exposures (years) under very similar conditions as the human exposures. In addition, it is anticipated that the information from these studies could relate to mechanism-generating hypotheses, particularly if initial plant investigations are followed up with more detailed analytical studies that investigate the mechanism of effect.

1

Introduction

This report contains the results and process of the National Research Council Committee to Assess the Potential Health Effects from Exposure to PAVE PAWS Low-Level Phased-Array Radiofrequency Energy. The PAVE PAWS radar system is located on Cape Cod, Massachusetts—near several populated residential areas and in a region known for its scenic beauty and ecological richness as well as being a popular tourist area. The purpose of this section is to describe the history of the facility and, in particular, the history of public concerns that have led to the committee's charge as well as to a number of other studies and investigations.

While the committee was established to perform an independent assessment of the possible biological and potential health effects of the PAVE PAWS phased-array radar system, and has a very specific statement of work (see below), the committee has taken very seriously the genesis for this study and, in particular, the specific concerns expressed by some members of the public.

HISTORY OF THE FACILITY

The PAVE PAWS phased-array radar system on Cape Cod was first conceived by the Joint Chiefs of Staff in 1972. The Raytheon Company was selected as the primary contractor on April 12, 1976, to build the facility. On March 17, 1976, the Air Force issued an environmental assessment of the PAVE PAWS radar system. For the purposes of that assessment, a power desity of 10 milliwatts per square centimeter (mW/cm^2) as averaged over any 6-minute period was established as a guideline for limiting occupational exposure. This guideline was provided by the U.S. Air Force School of Aerospace Medicine on the basis of existing Occupational Safety and Health Administration guidelines. In understanding the history of the community's concerns, it is important to note that these guidelines were not specified for non-occupational, continuous exposure. The Air Force's assessment considered a number of potentially detrimental scenarios to the environment due to the construction and operation of the radar and judged the impact to be minimal. It also noted considerations that were thought to offset the identified potential adverse environmental effects. The Air Force Environmental Assessment concluded in part that the proposed installation and operation of PAVE PAWS "is not a major federal action with significant adverse impact on the quality of the human environment." A report was also issued by a Department of Defense contractor, the Illinois Institute of Technology Research Institute, in May of 1976 that evaluated the impact of the proposed PAVE PAWS radar system on the electromagnetic environment at and near the proposed facility (IITRI 1976). Subsequently, on December 22, 1977, the Environmental Protection Agency (EPA) issued an environmental impact analysis of the PAVE PAWS system at the request of Massachusetts Representative Gerry E. Studds. For locations selected for evaluation that were beyond the boundaries of Otis AFB, the report concluded that "the predicted values of time-averaged power density are considered to be well below values that should have any health impact."

Following those studies, the Cape Cod Environmental Coalition, Inc. (CCEC), a citizens group, undertook court action in the U.S. District Court against various Air Force officials on March 3, 1978. The suit contended that the PAVE PAWS project was in violation of the National Environmental Policy Act of 1969 because an Environmental Impact Statement (EIS) had not been prepared. On October 31, 1978, the CCEC and the Air Force entered into an agreement temporarily suspending further litigation brought by the citizens group. Under the agreement, the Air Force was allowed to continue construction of the facility while it completed an EIS. The EIS was prepared by Stanford Research International under a contract with the Air Force in May of 1979.

Prior to completion of the EIS, the Air Force asked the National Academy of Sciences to perform two studies relevant to the EIS. One study was to focus on the safety measures inherent in the engineering design of the radar system and the second study was to focus on the extent of human exposure from the radar. These studies were completed in 1979. The results from the studies concluded that the "PAVE PAWS system may be anticipated to expose a limited number of members of the general public intermittently to low intensities of pulse-modulated microwave fields with maximal instantaneous intensities of 100 $\mu W/cm^2$ or less and time-averaged intensities lower by two orders of magnitude. There are no known irreversible effects of such exposure on either morbidity or mortality in humans or other species. Thus it is improbable that exposure will present any hazard to the public." However, the report went on to add, "In view of the known sensitivity of the mammalian central nervous system to electromagnetic fields, especially those modulated at brainwave frequencies, the possibility cannot be ruled out that exposure to PAVE PAWS radiation may have some effects on exposed people. Because these effects are still hypothetical, it is not feasible to assess their health implications. Such assessment will require additional research and surveillance and must be addressed in future evaluations of the potential exposure effects of PAVE PAWS and other high-power-output radar systems." Finally, the report recommended that additional research be conducted to clarify the possible effects of long-term exposure to the radar system. It is this recommendation that has formed the basis, in part, for some of the concerns expressed in the intervening 23 years by some local citizens prior to the start of this committee's current study.

Some of the concerns voiced by the public at the time of the 1979 study included concerns about possible thermal effects, disruption of implanted medical devices (such as pacemakers), and secondary radiation effects from improperly grounded structures exposed to the radar. In recent years, public concerns have shifted away from thermal effects of RF energy emitted by the radar and have centered on theories that relate to the waveform of the phased-array radar and concerns over the possible health ramifications from the propagation of the RF energy in tissue. A number of hypotheses have been advanced that propose possible mechanisms for health effects. In particular, some community members have made mention of studies done, and possible mechanisms of action advanced, by Dr. Richard Albanes.[1]

Dr. Albanese's concerns include potential biological effects on cells and tissues from:

- "Steep" rise-times in the PAVE-PAWS waveform,
- Overlapping wave fronts originating from multiple antennas on the PAVE PAWS radar, and
- The possibility of precursor formation in tissue.

Public concern over the potential for health effects from PAVE PAWS have been heightened by the Massachusetts Department of Public Health Cancer Registry reports, which for the last 10-plus years have reported on several excess cancers in the Cape Cod area, and, for certain cancers, on the upper Cape. Chapter 9 includes a summary and review of the available health studies and cancer registry information. Concern over the potential excess in cancers has been linked, for some members of the public, to potential environmental sources including PAVE PAWS.

Citizens expressed further concern over the PAVE PAWS radar system when

[1]Dr. Albanese is an Air Force employee who has raised concerns about the safety of the PAVE PAWS radar.

the Air Force announced in 1999 its proposed plans to upgrade the PAVE PAWS system. The Air Force has noted that the upgrades only involve system upgrades and do not involve physical expansion of the system. Following the Air Force's proposal, the Sandwich Board of Selectman and Senator Kennedy called for a site-specific EIS. The Sandwich Board of Selectman also requested that the Air Force perform laboratory studies of the effects of long-term exposure to PAVE PAWS and that a retrospective epidemiological study be done on Cape Cod to investigate the elevated cancer rates.

Since the time of this committee's inception, there have been citizens stating concerns on behalf of at least some members of the public about the absence of phased- array site-specific measurement data on PAVE PAWS from the Air Force for the past 23-plus years. That lack of data and failure to follow through on the recommendation made in the original NRC report have appeared to heighten the public's generalized concern and, for some, resulted in severe distrust of the Air Force. One such citizen's group is the Coalition for the Operation of PAVE PAWS Safely (COPPS). There has also been concern expressed by some members of the public that the group currently tasked with recommending and overseeing health or epidemiological studies—the local PAVE PAWS Public Health Steering Group (PPPHSG)—is not totally objective or independent of the Air Force, and thus, might not wholly represent some of the citizens who have been expressing concern. Distrust has been further elevated by a concern over potential information contained in the classified Air Force Environmental Health and Safety (EHS) program to which citizens requested that the current NRC committee be provided access. The committee's review of the EHS program and the committee's evaluation of the classified information are contained in Chapter 2 of this report.

Although laboratory studies of the effects of long-term exposure to PAVE PAWS waveforms have not been done, a retrospective epidemiological study and supporting power-density analysis have been commissioned by the PPPHSG. The power-density study was completed in May of 2004 and this committee anticipates an NRC review of the epidemiological study when it is completed in 2005. The committee's evaluation of the PPPHSG power-density study appears in Chapter 4 (Exposure Levels).

In summary, there are members of the Cape Cod public who have had longstanding concern over the potential health effects of the PAVE PAWS radar system. The absence of measured data and further specific studies since the facility was constructed and became operational have increased concerns and frustrated many residents.

ORIGIN OF PRESENT STUDY

In a January 11, 2001, letter from Senator Edward M. Kennedy to the Secretary of the Air Force, F. Whitten Peters, Kennedy asked that the Air Force fund an independent study through the National Research Council of the National Academies "to examine the health effects of the PAVE PAWS system." Kennedy further requested that this follow-on study (to the previous 1979 report) should address, at a minimum, the effects, if any, of the PAVE PAWS radar over the past two decades and should also examine the validity of using continuous-wave and pulsed non-ionizing radiation biological-effects data as surrogates for phased-array non-ionizing radiation biological-effects data. The following Statement of Task evolved out of discussions between the Massachusetts congressional delegation (Senators Kennedy and Kerry, and Congressman Delehunt), the Air Force, and the National Academies.

STATEMENT OF TASK

The committee will first determine whether continuous and pulsed radiofrequency (RF) energy research data are adequate for determining the biological and potential health effects of the PAVE PAWS phased-array system. This determination will be communicated to the sponsor as a letter report. If the research data from continuous and pulsed RF energies are considered to be applicable for the determination of possible health effects of phased-array RF energy, the committee will use this information to update the 1979 National Research Council analysis of the exposure levels and potential biological effects of the PAVE PAWS Radar System. If the data are not applicable, the committee will use other information that it determines to be relevant to phased-array health effects to update the 1979 report. In this update the committee will evaluate potential biological and health effects, evaluate exposures of the public to electromagnetic energy from the PAVE PAWS system, and make recommendations for the need for, and focus of, additional scientific studies to address continued scientific uncertainty related to health outcomes and exposure to low-level radiofrequency energy emitted by the PAVE PAWS Radar System. At the completion of its work, the committee will provide an update of the 1979 Research Council report that includes a discussion of:

1. The applicability of, and the level of uncertainty associated with, using data derived from cell, animal, and epidemiological studies employing continuous-wave exposure for evaluation of potential adverse health effects following phased-array exposures;
2. The extent of the exposure of the public to electromagnetic energy from the PAVE PAWS system;
3. Potential biological and health effects of the PAVE PAWS Radar System; and
4. Recommendations for appropriate follow-on study-design issues, including the strengths and limitations of the approaches suggested and the potential value of the proposed work.

Within the above specific scope, the committee has reviewed the information provided to us by members of the public, outside scientists, and the Air Force. The committee's scientific review process and deliberations in carrying out the statement of task have been performed with the highest level of scientific and professional integrity. It was also our intent that the process we used be made as transparent as possible. Furthermore, the draft report went through a rigorous independent peer review before the National Academies approved its issuance. The peer reviewers were anonymous to the committee until publication and the committee was required to respond to the satisfaction of the National Academies Report Review Committee before the draft was approved for release as a report of the National Academies. It is extremely difficult, if not impossible, to prove safety—it is, however, possible to examine whether there is a reasonable degree of certainty regarding the presence or absence of harm. In carrying out our statement of work and writing this report, the committee has attempted to evaluate the best available science and used this information and evaluation to address the public's concerns.

INFORMATION AVAILABLE TO THIS COMMITTEE

A large body of information on radiofrequency effects was available to this committee including RF information databases and recent reviews such as AGNIR (2001, 2003), Boice and McLaughlin (2002), GAO (2001), HCN (2000, 2002, 2003), ICNIRP 2001), Krewski and others (2001a, 2001b), NCRP (2002), NRPB (2004), SSI (2003), and Zmirou (2001). In addition to these databases and reviews, the committee had access to personal libraries accumulated during years of research by committee members on this and related subjects. To evaluate information the committee felt was most relevant to the PAVE PAWS issue, a table was constructed to focus the literature review on those components that the committee felt would be most useful to the present study. That table of criteria was published in a letter report delivered to the Air Force on November 15, 2002, and is included in this report as Appendix A.

Secondly, the Air Force made a PAVE PAWS waveform study available to the committee (AFRL 2003) and the PAVE PAWS Public Health Steering Group made a PAVE PAWS power-density study available to the committee through their contractor, Broadcast Signal Laboratories (BSL 2004).

Finally, a large amount of information was made available to the committee by citizens and other individuals interested in the Academies PAVE PAWS study.

UNCERTAINTIES ASSOCIATED WITH THE EVALUATION OF PAVE PAWS HEALTH EFFECTS

The Statement of Task asks the committee to discuss the level of uncertainty associated with using data derived from cell, animal, and epidemiological studies employing continuous-wave exposure for evaluation of potential adverse health effects following phased-array exposures. Some of the uncertainties associated with using data derived from studies employing continuous-wave or pulsed-wave exposure for evaluation of potential adverse health effects following phased-array exposures include:

- The extent to which pulse shape, repetition rate, and carrier frequency affect biological systems;
- The extent to which power levels and exposure duration for a given waveform affect biological systems; and
- Identification of the relevant biological effects and how they relate to health outcomes, including the state of the biological system and the pathways by which biological changes occur. Biological test systems may not closely approximate the physiological state in humans, and human physiological conditions vary widely. The approaches to extrapolate from biological effects to health effects are frequently not well defined.

The committee emphasizes that uncertainties are associated with all scientific information and analyses and these uncertainties and others were taken into account in reaching the committee's conclusions and recommendations.

REFERENCES

AFRL (The Air Force Research Laboratory). 2003. Phase IV—Time-domain Waveform Characterization Measurements of the PAVE PAWS Radar. Air Force Research Laboratory, Kirtland Air Force Base, Albuquerque, NM.

AGNIR (Advisory Group on Non-Ionising Radiation). 2001. Possible Health Effects from Terrestrial Trunked Radio (TETRA). Report of an Advisory Group on Non-ionising Radiation. Docs NRPB, (12)2 1-53. Available from http://www.nrpb.org/publications/documents_of_nrpb/abstracts/absd12-2.htm.

AGNIR. 2003. Health Effects from Radiofrequency Electromagnetic Fields. Report of an independent Advisory Group on Non-ionising Radiation. Docs NRPB, 14 1-177. Available from http://www.nrpb.org/publications/documents_of_nrpb/abstracts/absd12-4.htm.

Boice, J.D. and J.K. McLaughlin. 2002. Epidemiologic Studies of Cellular Telephones and Cancer Risk—A Review. SSI Report 2002:16. Stockholm: Statens Strålskyddsinstitut. Available from http://www.ssi.se/english/Press_release_rapp2001_16.html.

BSL (Broadcast Signal Lab). 2004. A Survey of Radio Frequency Energy Field Emissions from the Cape Cod Air Force Station PAVE PAWS Radar Facility: Final Test Report, prepared for PAVE PAWS Public Health Steering Group by Broadcast Signal Lab, Medfield MA.

GAO (General Acct. Office). 2001. Telecommunications: Research and Regulatory Efforts on Mobile Phone Health Issues. Report to Congressional Requesters. GAO-01-545. Washington DC: United States General Accounting Office. Available from http://www.gao.gov/new.items/d01545.pdf.

HCN (Health Council of the Netherlands). 2000. Electromagnetic Fields Committee. GSM Base Stations. Publication Number 2000/16E. The Hague: Health Council of the Netherlands.

HCN. 2002. Electromagnetic Fields Committee. Mobile Telephones: An Evaluation of Health Effects. Publication Number 2002/01E. The Hague: Health Council of the Netherlands.

HCN. 2003. Health Effects of Exposure to Radiofrequency Electromagnetic Fields: Recommendations for Research. Publication Number 2003/03. The Hague: Health Council of the Netherlands.

ICNIRP (International Commission on Non-ionizing Radiation Protection). 2001. Biological Effects, Health Consequences and Standards for Pulsed Radiofrequency Fields in Proceedings of the International Seminar on Biological Effects, Health Consequences and Standards for Pulsed Radiofrequency Fields, Erice, Italy, November 21-25, 1999, R. Matthes, J.H. Bernhardt, and M.H. Repacholi eds. International Commission on Non-ionizing Radiation Protection ISBN 3934994008.

IITRI (Illinois Institute of Technology Research Institute). 1976. EMC Analysis of Otis AFB PAVE PAWS Site. Electromagnetic Compatibiliy Analysis Center ECAC-PR-76-020, Annapolis, MD.

Krewski, D., C.V. Byus, B.W. Glickman, W.G. Lotz, R. Mandeville, M.L. McBride, F.S. Prato, and D.F. Weaver. 2001a. Potential health risks of radiofrequency fields from wireless telecommunication devices. J Toxicol Env Heal B 4:1-143.

Krewski, D., C.V. Byus, B.W. Glickman, W.G. Lotz, R. Mandeville, M.L. McBride, F.S. Prato, and D.F. Weaver. 2001b. Recent advances in research on radiofrequency fields and health. J Toxicol Env Heal B 4:145-159.

NCRP (National Council on Radiation Protection and Measurements). 2002. Biological Effects of Modulated Radiofrequency Fields. NRCP Commentary No 18. Bethesda, Maryland: National Council on Radiation Protection and Measurements.

NRC (National Research Council). 1979. Analysis of the Exposure Levels and Potential Biologic Effects of the PAVE PAWS Radar System. Washington, DC: National Academy Press.

NRPB (National Radiological Protection Board). 2004. Review of the Scientific Evidence for Limiting Exposure to Electromagnetic Fields (0-300 GHz). Docs NRPB 15 (3) 1-215. Available from http://www.nrpb.org/publications/documents_of_nrpb/abstracts/absd15-3.htm.

SSI (Swedish Radiation Protection Institute). 2003. Recent Research on Mobile Telephony and Cancer and Other Selected Biological Effects: First Annual Report from SSI's Independent Expert Group on Electromagnetic Fields. Available from http://www.ssi.se/english/EMF_exp_Eng_2003.pdf.

Zmirou, D. 2001. Mobile Phones, Their Base Stations, and Health. Report to the French Health Directorate. France, Direction Générale de la Santé. Available from http://www.sante.gouv.fr. Summary in English available from http://www.sante.gouv.fr/htm/dossiers/telephon_mobil/pdf/teleph_uk.doc.

2

Assessment of Classified Research Relevant to PAVE PAWS

INTRODUCTION

Because radars are an important feature in the operation of the military in times of peace and war and because microwave and laser systems have been developed as strategic weapons to deliver energy as well as systems for guidance or surveillance, one would expect that the U.S. Air Force and perhaps other branches of the military might have conducted research to explore effects of microwaves on biological systems. A major source of that type of research has been a responsibility of the Directed Energy Bioeffects Division in the Human Effects Division (HED) of the Air Force Research Laboratories (AFRL) at Brooks Air Force Base in San Antonio, Texas. For obvious reasons, some of that work has been classified by the U.S. government, while other results in biological systems have been published in the open scientific literature or in unclassified U.S. Department of Defense documents. The congressionally mandated study conducted by the NRC committee included a charge to examine and assess all classified national security data that might be relevant to PAVE PAWS.

VISIT DETAILS

In order to ascertain whether research by the U.S. Air Force had produced any evidence for biological effects that might relate to potential health effects of electromagnetic radiation with characteristics relevant to the radiations produced by the PAVE PAWS beam, members of the NRC committee with appropriate personnel security clearances and expertise in the engineering and biological disciplines were asked to review the Air Force's classified research and results. To do that part of the charge, two site visits were conducted. The first was a preliminary visit conducted on April 25, 2002, in which the goal was to examine the reports and documents summarizing all of the classified research results that might be relevant to the PAVE PAWS study and to determine whether additional expertise and a second visit would be necessary to review the Air Force's results of classified research. The Director of the Board on Radiation Effects Research and two committee members, one with biological expertise and one with physics expertise, examined the classified reports and data summaries in the Directed Energy Bioeffects Division of the laboratories at Brooks AFB. The site-visit team asked the Air Force to provide all reports and data summaries for all electromagnetic radiation research related to the PAVE PAWS beam characteristics. This visit included an extensive presentation by Dr. Albanese, in addition to presentations by other scientists, including Air Force personnel and civilians who were associated and familiar with the Air Force's research activities related to electromagnetic radiation interactions with biological systems. The team was hosted in Building 1162 (Tejeda Laboratory) at Brooks Air Force Base by Richard L. Miller, Ph.D., Chief, Directed Energy Bioeffects Division (AFRL/HED). Also present were Major Lester Ogawa, Dr. Johnathan Kiel, the Senior Scientist in Electromagnetic Radiation Effects, and Dr. Walter Rogers. At the time of the NRC committee's visit, Dr. Rogers was a Research Electrophysiologist for Veridian, but he had been the key investigator on the Electromagnetic Health and Safety program report when he was an employee for Southwest Research Institute. Part of the review included a discussion and answer session with scientists and Air Force administrative staff and part of the review included an executive session in which the NRC site-visit team examined reports and data summaries.

The second site visit was conducted on January 30-31, 2003. For the second visit, two additional committee members with appropriate personnel security clearance and scientific expertise (the additional expertise requirement was based on the review conducted by the first review team) were added to the team. In addition, a physician with appropriate personnel security clearance was included as an unpaid consultant. A total of four committee members, plus an M.D. consultant (a member of the Institute of Medicine of the National Academies) and the NRC board director, reviewed the classified information provided during the two-day visit.

Prior to the site-visit team's arrival, the full committee's members developed a series of questions that they instructed the review team to ask of the Air Force during their visit. Those questions included:

1. Where did the parameter "1 volt/meter/nanosecond rise time" come from?

2. Does the Air Force have any evidence for actual measurements of precursors?

3. Does the Department of Defense have any other data (molecular, cellular, or animal) that are relevant to the PAVE PAWS exposures (i.e., other experimen-

tal results known to the Air Force from experiments conducted using relevant wavelengths and power densities that are below thermal thresholds)?

In addition to focusing on a discussion of the Air Force's answers to the above questions, a major focus of the review was a final report of the U.S. Air Force's Electromagnetic Health and Safety (EHS) research program. The results of the Air Force-sponsored studies conducted over a period of approximately 15 years (from the 1980s to 1996) were reviewed from the perspective of whether any of the results obtained are relevant to exposure of humans to the PAVE PAWS system or informative about effects that might relate to potential human health effects from exposures to PAVE PAWS. As stated in the final report provided by the Air Force, the thrust of the EHS program is summed up by the unclassified title *Biological Effects of Exposure to Ultra-wideband Electromagnetic Energy (U)* (AFRL-HEX 2002). Copies of the final classified report were provided to each review-team member for use during the two-day visit.

In its review of the classified data, the review team had to rely on the responsible Air Force personnel to provide access to all data as requested by the committee. The Air Force administrative personnel were very cooperative in making the classified materials of the EHS program available to the committee. In this particular review, the committee has no evidence to suggest that summaries of data obtained by the Air Force in research related to the EHS research program were withheld from the committee. Because of the large volume of data obtained over several years in this multimillion-dollar research program, the reviewed materials were primarily summary and tabulated material. Information provided in the summary reports did not indicate the existence of biological effects in the accumulated data that are relative to PAVE PAWS exposures. It was helpful that the review team was able to meet separately with Dr. Albanese, given that he had raised his personal concerns on several occasions to the full NRC committee and to the site-visit team. Dr. Albanese was able to observe the materials and briefings that were provided to the review team, and was asked specifically whether there was any additional information that should be reviewed by the team. The two suggestions made by Dr. Albanese were followed up by the team: first, the suggestion that the U.S. Navy might have conducted some relevant research; and second, the suggestion that contractors at other sites conducted experiments that might be relevant (see Chapter 5, Annex 5-2). Finally, members of the review team and the full NRC committee feel that it would be prudent for the Air Force to declassify as much of the classified EHS material as possible in the interest of improving the public's confidence in the U.S. Air Force's disclosure, even though the review team saw no evidence of results that would change the NRC committee's conclusions or recommendations related to its assessment.

The NRC's review team members produced an unclassified summary and presented it to the full committee at its meeting the following month in Washington, DC, February 10-11, 2003. Based on the report by the review-team members, and following discussion by the full committee, the following conclusions and findings were made:

CONCLUSIONS AND FINDINGS

1. As best as could be determined from materials provided at our request, the Air Force did not design or conduct any classified studies addressing long-term exposure effects experiments directly relevant to the PAVE PAWS exposure conditions and therefore did not respond to the recommendations in the 1979 NRC Report.[1]

2. However, the Air Force has completed a series of studies using high-power, short-duration RF exposures that include energy in the frequency range and average power of PAVE PAWS.

3. There are no classified experimental studies in the EHS program relating to carcinogenesis.

Experimental results reviewed by the site-visit team, while not generally statistically significant or at exposure conditions representative of PAVE PAWS, suggest that stimulation of cardiac or skeletal muscle might be demonstrated at high power levels and sub-second exposures. It was somewhat disappointing to the review team that some of the data summaries could not be easily traced back to certain key experimental characteristics.

The 1 volt/meter/nanosecond slope appears to be based on Dr. Albanese's theoretical modeling and measurements in "biologically relevant" models that have been declassified.

The visit did not reveal any additional measurements of, or demonstration of, precursors in biological or model systems.

The review team learned that the U.S. Navy has conducted studies of transient microwave propagation in ocean media. The Navy was contacted and the committee received an unclassified report from Naval Surface Weapons Center (NSWC) (see Chapter 5 this report).

[1]The 1979 report recommended: "Additional research is recommended to clarify further the possible effects of long-term exposure to microwave radiation at low power densities. . . . In view of the known sensitivity of the mammalian central nervous system to electromagnetic fields, especially those modulated at brainwave frequencies, the possibility cannot be ruled out that exposure to PAVE PAWS radiation may have some effects on exposed people. Because these effects are still hypothetical, it is not feasible to access their health implications. Such assessment will require additional research and surveillance and must be addressed in future evaluations of the potential exposure effects of PAVE PAWS and other high-power-output radar systems."

REFERENCES

NRC (National Research Council). 1979. Analysis of the Exposure Levels and Potential Biologic Effects of the PAVE PAWS Radar System. Washington, DC: National Academy Press.

AFRL-HEX (Air Force Research Laboratory–Health Effects Classification). 2002. Biological Effects of Exposure to Ultra-wideband Electromagnetic Energy (U). AFRL-HEX classified report. Brooks Air Force Base, San Antonio, TX.

3

Physical Mechanisms for RF Effects on Biological Systems

There is a long chain of events that must be followed to go from fundamental physical interactions between electromagnetic fields and molecules or ions to the production of changes in biologic systems that could lead to adverse human health effects. This chain of events has yet to be described in detail. This section will outline a small portion of those events, beginning with the forces of electric and magnetic fields on electrons, ions, atoms, and molecules, and changes in their energy, configuration, or orientation. These changes in turn can lead to further changes in chemical reaction rates and in the binding of molecules to membranes. This progression can then lead to changes in cell activity that in turn can affect the biology of the organism. A subset of these biologic changes can lead to adverse health effects. The body has many feedback processes so that many biologic deviations from the norm are corrected before they become adverse health effects. This section of the report will focus on the physics involved in the initial portion of this chain of events.

The depth of RF penetration is dependent on the conductivity and dielectric constant of the biological material; usually in the range of 3 to 4 cm at 433 MHz for typical biological materials and ranges up to 16 cm in bone. Thus one can expect a significant fraction of the field to penetrate into the body.

Maxwell's equations, the Lorentz force law, and the electrical characteristics of the material in question describe the basic interaction between electromagnetic waves and that material. The electric field is defined so that the force on a charged particle is given by the product of the charge, q, and the electric field, E. Similarly the force exerted by the magnetic flux density, B, is given by the vector product of the velocity, v, of a charged particle and the magnetic flux density. The vector product means that this force is at right angles to both the velocity of the particle and B or it is perpendicular to the plane formed by these vectors. Given the value of the fields E and B, one can calculate the forces, F, being exerted on an electron, an ion, or a charged molecule. These relations can be expressed by the following formula:

$$F = q(E + v \times B).$$

From these forces and the equations of motion, one can calculate changes in motion and kinetic energy of the particles as a result of the application of the fields as a function of time and space. It should be noted that at the PAVE PAWS radar frequency of 420 MHz, the direction of the force reverses every half cycle or every 1.2×10^{-9} seconds. Therefore, the average displacement of these charged particles in a uniform field is zero. Thus the dominant result of a uniform field is heating. An average incident power density of 1 $\mu W/cm^2$ on a standing man would lead to a specific absorption rate, SAR, of approximately 5×10^{-5} W/kg. This is about four orders of magnitude lower than the metabolic-energy generation rate of a resting man.

In a non-uniform field the gradient of these fields can induce a directional force in atoms and molecules. If for example, the electric field induces a dipole moment that oscillates with the field, then the gradient of the field will exert a force on the particle that is in a constant direction. Thus the gradient of the field can lead to a drift current density, J, that is given by the formula:

$$J = N \mu\alpha V(E \bullet \nabla)E,$$

where α is the polarizability, N is the density of the particles, V is the volume of the ion or molecule, and μ is the mobility. For the fields under consideration, this induced drift current density will be very small and is expected to be very small with respect to the diffusion current and the drift currents associated with the fields that occur naturally around cells associated with biological activity that are on the order of microamps per centimeter squared.

The applied fields can also change the state or energy of the bound electrons in an atom or molecule. For weak fields these changes are associated with the absorption of a quantum of energy from the RF field. The amount of energy, W, associated with a single quantum of the RF field is given by the formula:

$$W = hf,$$

where h is Planck's constant and f is the frequency. The quantum of energy associated with a photon of microwave energy is about 10^{-5} times smaller than the photon of energy associated with room temperature radiation. The thermal radiation has its maximum energy in the infrared. This thermal background radiation nearly equalizes the population of low energy states that are separated by energies corresponding to a microwave photon at thermal equilibrium so that many quantum effects are completely masked by the thermal background radiation.

One area in which this may not be true is associated with excited states where most of the energy is supplied by another process such as a chemical reaction or ultraviolet radiation. For these excited states, spin selection rules may control the rate at which they decay or react with other materials and the population of molecules in these states may be changed by RF radiation. Examples where these processes may be important are free radicals in which relatively low levels of RF power have been shown to change the absorption spectra. In the 1-80 MHz region of the spectra, and at field strengths of 0.1 to 0.5 mT, Stass and others (2000) have shown a magnetic field effect on the photochemical reaction of anthracene-d_{10} with 1,3-dicyano-benzene in a cyclohexanol/acetonitrile solution corresponding to changes in the free radical life times. These transitions are associated with hyperfine spectra of the molecules.

In general, the fields are attenuated as they propagate through the tissue. It is a relatively complicated problem to estimate how strong the fields will be after they go through the skin and other anatomy taking into account the geometry of the body and the differing electrical properties of the skin, bone, fat, and other tissues to find the field strengths at the site of interest for a given biological effect. Tables for the electrical properties of many tissues, are given by Gabriel, Lau, and Corthout as a function of frequency in three papers . Using this kind of data, numerical models have been used to calculate the field distributions in the head and the body in various positions (Hagmann and Gandhi 1979; Jensen and Samii 1995; Iskander and others 2000). For a review of computational methods for computing field distributions see the *Handbook of Biological Effects of Electromagnetic Fields*, Chapter 9 (Lin and Gandhi 1996). In general, the fields will be weaker the farther away the biological site of interest is from the surface facing the source. In brain tissue the attenuation coefficient is 31.1 m^{-1} and the depth of penetration is about 3.2 cm. The fields in membranes and other low water content material will be larger than in the high water content material by the ratio of the dielectric constants. This is about a factor of 20 at 420 MHz for membranes in a fluid.

The effect of the RF fields on the biological system may take place by changing chemical reaction rates or the binding of molecules to a membrane surface. This could occur in at least five ways (Barnes 1996). First, it may affect the transport of ions or charged molecules and thus the probability of the two particles coming close enough to each other to react. Second, it may affect the energy with which they collide. Third, it may affect the orientation or configuration of the colliding particles. Fourth, it may change the energy state of one of the molecules. Fifth, it may affect the average temperature of the environment. Of these effects, only those related to changes in the average temperature are well studied and are generally accepted by the scientific community at large as described in a review article by Adair (2003).

Effects that are currently being studied include changes in the molecular configuration of large biological molecules as a result of the application of RF fields, dielectrophoresis or the effects of the gradients of the fields on the transport of molecules in the vicinity of membranes, and the effects of RF fields on free radical lifetimes. For dielectrophoresis to be important the resulting current must be a significant fraction of the natural current of the same material. Initial results of the studies of these mechanisms indicate that to be important, the field strengths need to be sufficient so that the energy absorbed from the RF fields is a reasonable fraction of the thermal energy, kT.

Other physical changes that have been suggested include changes in the diffusion constants, (Seto and Hsieh 1976) and rectification of the electric field by membranes (CRC 1996). For these to occur, the signals must be large enough for non-linearities to become significant. Non-linearities are expected to be most important in biological systems with gain. At low frequencies, one such system is the cardiac pacemaker where_cells were shown to have nonlinear effects on the oscillation frequency on the order of a hundred microvolts (CRC 1996). However, at frequencies above 10 MHz the membrane capacity shorts out the applied field and the field reverses direction so rapidly that ions cannot transit a membrane. Measurements of RF field range applied to cell membranes have not shown any rectification (Pickard and Barsoum 1981).

An important problem is determining the minimum signal that a biological system can detect in the presence of noise. For reliable communications signal-to-noise ratios of greater than one are usually required and typical values are one hundred-to-one and one thousand-to-one. Sources of noise include thermal noise, shot noise, 1/f noise, and the electrical signals generated by other parts of the biological system. At low frequencies, the electrical signals from muscle activity of the heart are usually the largest source of electrical noise. In biological systems, repetitive pulses that have a pattern that can be distinguished from the noise are usually required to initiate important changes or to put information into memory. These signals often have both space and time coherence. For signals that are coherent in space, the signal-to-noise ratio grows as the square root of the number of events in parallel with respect to random noise. For signals that are coherent in time one gets a similar increase in signal-to-noise ratio with the square root of the length of time for which the signal is applied (Weaver and Astumian 1990).

At 420 MHz, the thermal noise is expected to be the largest source of noise. The noise from this source leads to mean-squared average-voltage fluctuation across a membrane that is given by

$$\langle V^2 \rangle = 4kT\Delta fR,$$

where k is Boltzmann's constant, T is the absolute temperature, Δf is the bandwidth, and R is the resistance.

The value of the resistance and the effective bandwidth of cell membranes will depend on the geometry of the cell. Additionally, because cells are electrically coupled to each other, the effective values for the resistance and capacitance of a cell will depend on its environment and thus the noise voltage will also be dependent on the geometry of the cells. For an externally applied voltage, the voltage across a particular cell is dependent on the geometry of the cell and the structure of the surrounding tissue. Typical models for a cell membrane would consist of resistor and capacitor in parallel. Another resistor and capacitor in parallel would model the fluid portion of the cell and the applied field would extend across a large number of cells in series. At high frequencies, this would behave like a capacitive voltage divider so that the applied voltage to a given membrane would be roughly equal to the applied field divided by twice the number of cells per unit distance.

At low frequencies the currents through a cell membrane are nonlinear functions of the applied voltage. These nonlinearities are such that cell membranes in nerve cells can behave as poor rectifiers with a typical efficiency of about 0.1%. At radiofrequencies, this efficiency has typically not been measured and there is at least one proposal that is outstanding to make measurements of this kind.

A single molecular event can be amplified by a variety of means. For example, the binding of a single neural transmitter at a synaptic junction can lead to the release of thousands of calcium ions that in turn become a part of the signal that is used to excite the next synapse. It typically takes approximately 20 dendritic inputs to a summing junction to fire an axon. Repetitive stimulation can lead to persistent changes that either increase or decrease the threshold for firing. Stochastic resonance is another means of amplification that can lead to an increase in the signal-to-noise ratio. In these processes a small periodic signal in a nonlinear system can be amplified by extracting energy from the noise. Gains on the order of 100 and similar improvements in the signal-to-noise ratios have been shown for physical systems, and stochastic resonance has been shown to be one method for improving the sensitivity of biological systems to weak signals (Gammaitoni and others 1998).

SUMMARY

There are a number of possible mechanisms and pathways by which electric and magnetic fields could lead to biological changes at high-power exposure levels. However, at this time, the committee does not know of a physical mechanism that has been shown to change biological processes at the field-strength levels associated with exposures to the PAVE PAWS radar.

REFERENCES

Adair, R.K. 2003. Biophysical limits on athermal effects of RF and microwave radiation. Bioelectromagnetics 24:39-48.

Barnes, F.S. 1996. Effect of electromagnetic fields on the rate of chemical reactions. Biophysics 41:801-808.

CRC. 1996. Interaction of DC and ELF electric fields with biological materials and systems. In: The CRC Handbook on Biological Effects of Electromagnetic Fields, C. Polk and E. Postow, eds. Boca Raton, FL: CRC Press.

Gabriel, C.S., C. Gabriel, and E. Corthout. 1996a. The dielectric properties of biological tissues. I. Literature survey. Phys Med Biol 41:2231-2249.

Gabriel, C.S., R.W. Lau, and C. Gabriel. 1996b. The dielectric properties of biological tissues. II. Measurements in the frequency range 10 Hz to 20 GHz. Phys Med Biol 41:2251-2269.

Gabriel, S., R.W. Lau, and C. Gabriel. 1996c. The dielectric properties of biological tissues: III. Parametric models for the dielectric spectrum of tissues. Phys Med Biol 41:2271- 2293.

Gammaitoni, L, P. Hänggi, P. Jung, and F. Marchesoni. 1998. Stochastic resonance. Rev Mod Phys 70:223-287.

Hagmann, M.J. and O.P. Gandhi. 1979. Numerical calculation of electromagnetic energy deposition in models of man with grounding and reflector effects. Radio Science 14:23-29.

Iskander, M., Z. Yun, and R. Quintero-Illera. 2000. Polarization and human body effects on the microwave absorption in a human head exposed to radiation from handheld devices. IEEE Trans MTT 48(11):1979-1987.

Jensen, M. and Y. Rahmat Samii. 1995. EM interaction and handset antennas and a human in personal communications. P IEEE 83:7-17.

Lin, J. and O. Gandhi. 1996. Computer methods for predicting field intensity. In Handbook of Biological Effects of Magnetic Fields, edited by Charles Polk and Elliot Postow. CRC Press.

Pickard, W.F. and Y.H. Barsoum. 1981. Radio-frequency bioeffects at the membrane level: separation of thermal and athermal contributions in Characeae. J Membrane Biol 61:39.

Seto, Y.J. and S.T. Hsieh. 1976. Biotechnol Bioeng 18:813.

Stass, D.V., J.R. Woodward, C.R. Timmel, P.J. Hore, and K.A. McLauchlan. 2000. Radio frequency magnetic field effects on chemical reaction yields. Chem Phys Lett 320:15-22.

Weaver, J.C. and R.D. Astumian. 1990. The response of living cells to very weak electric fields: The thermal noise limit level. Science 247:459-461.

4

Exposure Levels

OVERVIEW

This chapter summarizes the current data on exposure levels produced by PAVE PAWS. It begins with an overview of the operating characteristics of the radar and then discusses the existing exposure data measured in terms of peak and average power density (in $\mu W/cm^2$). Recent measurements of PAVE PAWS power density recorded by census tract obtained as part of the current initiative to re-examine the potential health effects of exposure to low-level phased-array RF energy are included. Discussion also considers exposure levels of the general population to other broadcast radiation sources of comparable spectral content.

PAVE PAWS OPERATION

The environmental impact and safety questions related to human health for radar installations (and other broadcast sources) have traditionally been characterized in terms of the power density produced as a function of time and location within the surrounding population (often neglecting details of the local terrain). Power density remains an important metric for safe operation of a radar in a populated area. As such, the primary physical characteristics of PAVE PAWS that determine its power-density distribution (e.g., its beam attributes) are briefly reviewed (detailed descriptions of the full operational behavior of PAVE PAWS have appeared in prior reports [NRC 1979a, b]). Historically, power levels below the limits known to cause thermal effects in tissue have been deemed acceptable. Current safety standards, which have been in place in this country for a number of years, continue to evolve and are periodically revised, but are essentially based on this threshold (e.g., IEEE Standard C95.1-1999). However, concerns have recently been raised regarding the safety of exposure to phased-array radiofrequency energy. Specifically, two questions have been articulated. The first is whether or not the human body responds differently (and detrimentally) to a radar beam formed from a large number of individual antenna elements radiating waveforms that are slightly shifted in time (to form a beam whose direction can be electronically scanned by controlling the phasing between elements) relative to a single source generating an equivalent beam (in terms of radar function) that is mechanically redirected. The second is whether or not the transient characteristics of the waveforms generated by PAVE PAWS are sufficient to produce so-called *precursors*—spectral components of the composite signal created through interactions with the human body that have been hypothesized to propagate to distances into tissue extending beyond that expected from the primary (signal) frequency comprising the PAVE PAWS emission. As a result, the summary of PAVE PAWS operation presented here will focus on its beam-forming capabilities. The characteristics of the time-domain signal measured from PAVE PAWS and the conceptual and experimental basis for the existence/nonexistence of precursors generated in tissue by PAVE PAWS exposure will be developed more fully in Chapter 5.

THE BEAM CHARACTERISTICS

PAVE PAWS consists of two planar arrays of active antenna elements (the centers of each array face are pointed 20° above horizontal and 120° apart in azimuth) that are phased to radiate a directed narrow (main) beam (2° in width, broadside) in short, high-energy bursts (pulses) containing approximately 90% of the total transmitted power at the intended RF frequencies (any one of 24 discrete frequencies between 420 to 450 MHz). The remaining energy is distributed in sidelobes to the main beam which form clusters of surrounding auxiliary beams that transmit significantly reduced intensities pointed in directions fanning out from the orientation of the main beam with respect to the horizontal. For example, the first and second sidelobes radiate powers that are no more than 1% and 0.1% of the main beam, respectively are directed approximately 4° (1st sidelobe) and 6° (2nd sidelobe) off the central axis of the main beam (horizontally). Given the local topography of the Cape Cod PAVE PAWS installation (which slopes away from the radar at 1° or more below horizontal), it is primarily the secondary sidelobes that intersect the ground when the main beam is pointed at its lowest (3° above horizontal) elevation.

Transmitted Power

Each face of the PAVE PAWS radar consists of a regular grid of 2677 antennas of which 885 are inactive, leaving 1792 active that are directly connected to individual solid-state transmitters radiating about 320 watts at peak power. The peak transmitted power of PAVE PAWS is rated at 580 kW (320 W × 1792), since typically only one of the two antenna faces radiates at a time. As described in more detail below, the sequence of pulses radiated depends on the function that the radar is performing at the time. Controls built into the system limit the rate at which pulses are transmitted so that the fraction of total time during which power is radiated does not exceed 25%, making the average transmitted power less than 145 kW (580 kW × 0.25). Correspondingly, the average radiated power in the first and second sidelobes is approximately 1.45 kW and 145 W, respectively.

Waveform Generation

PAVE PAWS operates by transmitting pulses of radiated power interspersed with periods of time dedicated to signal reception, the details of which depend on the radar's immediate functional task (e.g., searching, tracking), making the type of pulse or pulse burst that is emitted, and the timing between bursts, complex and dynamic. Further, the variability in the pulse pattern is coupled to directional changes and small pulse-to-pulse frequency shifts in the main-beam transmissions that create sidelobe radiation patterns that are constantly changing at any given location on the ground. It is, therefore, difficult to quantify precisely exposure conditions over time because the patterns of pulses emitted by PAVE PAWS depend on the scanning directions being searched by the main beam along with the number and location of targets being tracked at a particular instant in time.

The operation of PAVE PAWS is synchronized to a 54 msec cycle time in which one or more pulses may be transmitted in each of several directions at slightly different frequencies. Seventeen (17) consecutive 54 msec cycles are devoted to radar function with an 18th used for self-testing. Within a cycle, pulse widths ranging from 0.25 to 16 msec may be generated. Within some pulses, the main transmission frequency is varied (modulated or "chirped") by 2 MHz or less depending on radar function. Hence, the maximum possible duty cycle (fraction of time that power is transmitted) during a single pulse repetition interval is 30% (16/54 msec) and 28% over 18 consecutive cycles (17/18 × 30%), although on average the maximum duty cycle does not exceed 25%.

Although the waveform modulation scheme is complicated by the multi-task operation of PAVE PAWS, the modulation of transmitted power experienced at a fixed location on the ground has been estimated based on the assumption that the radar is functioning in an enhanced search mode (where the main beam targets 120 different azimuthal locations at 3° above the horizon every 2.5 seconds) (NRC 1979b). In this analysis, the signal (assumed to be a nominal 435 MHz carrier) was considered to be modulated both by the pulse pattern of the burst transmissions and by the scanning of the sidelobes as the main beam is repositioned. The power spectrum of the envelope of the enhanced search mode pulse train with a repetitive period of 54 msec contains a small amount of energy at 18.5 Hz (1/54 msec) and its harmonics. The periodic interruptions (once every 18 cycles for antenna self-testing) in transmissions create small sidebands about these frequencies spaced approximately every 1 Hz (18×54 msec = 0.972 sec; 1/0.972 = 1.03 Hz). The 1979 NRC report (NRC 1979b) incorrectly states that 30% of the total power is at zero frequency. This finding was based on the envelope of the pulses. In reality, each pulse is filled with oscillations at the radar frequency (420-450 MHz); each pulse has a very small DC component. Because of this, the 18 msec maintenance sequence used every 972 msec (one of 18) also has very small DC components.

As noted in NRC 1979b, continuous operation in enhanced search mode is not probable in practice, meaning that the 18.5 Hz peak would be reduced with more power distributed around its overtones (37 Hz, 74 Hz, 148 Hz) resulting from the factors of two reduction in pulse widths that occur with the variation in transmission bursts. Additional sidebands would be present as well due to the quasi-periodic recurrence of the shorter interval pulses occurring within the 54 msec repetition rate. Further spectral spreading results from the pulse-to-pulse sampling of the antenna pattern sidelobes experienced at the fixed location on the ground as the direction of the main beam is changed. This net effect was estimated to create a uniform loss (~6.6 dB) across the band with further spectral spreading into the sidebands of the pulse-train envelop spectrum leading to a distribution dominated by sidelobe power in the 0-3 Hz band with less than 1% of the total in the 15-20 Hz range (NRC 1979b). The report describes a sample strip-chart recording of power measurements in the field ,which showed a 2 Hz fluctuation with a peak amplitude of approximately 0.4 $\mu W/cm^2$ on which spikes in amplitude of about 1.4 $\mu W/cm^2$ were superposed with a regular period of 0.4 Hz. These findings tend to confirm the estimate that the modulation of the sidelobe power is a few hertz or less on the ground.

PAVE PAWS POWER-DENSITY ESTIMATES AND MEASUREMENTS

Exposure levels on the ground from PAVE PAWS have been estimated and measured a number of times at various locations under several different assumptions and conditions. Six sources of such data are reviewed and summarized here: (1) the 1979 NRC (National Research Council) Engineering Panel report on Radiation Intensity of the PAVE PAWS Radar System (NRC 1979b); (2) the 1979 NRC report on Exposure Levels and Potential Biological Effects of the PAVE PAWS Radar System (NRC 1979a); (3) the 1986 Engineering Report (#86-33) issued by the 1839 Engineering Installation Group (EIG) from the Engineering Division at Keesler Air Force Base entitled "Radio Frequency Radiation Survey" for the AN/FPS-115 PAVE PAWS Radar (EIG 1986); (4) the 2000 report by Mitre Corporation on "RF Power Density Exposure at Ground Level for the PAVE PAWS Radar at Cape Cod—Questions and Answers" (MITRE 2000); (5) the Broadcast Signal Laboratory (BSL) final test report "A Survey of Radio Frequency Energy Field Emissions from the Cape Cod Air Force Station PAVE PAWS Radar Facility" issued in April of 2004 (BSL 2004b); and (6) the power density measurement data analysis from the Coalition to Operate PAVE PAWS Safely (COPPS) provided in May 2004.

1979 NRC Engineering Panel Report

The 1979 National Research Council Engineering Panel reviewed measurement data obtained from June 1978 through August 1978 as well as other recordings presented to that committee in September 1978 (NRC 1979b). These measurements were obtained with the radar operating in the enhanced search mode at a fixed carrier frequency with 20% duty cycle and were corrected for equivalency to measurements at 1 km distance from the radar with a 25% duty cycle (for comparison purposes). Data from 4 station locations ranging from 1600 to 3900 feet from the radar (at 0° to 63° in azimuth from boresight) resulted in average power densities of 0.38 to 3.26 mW/cm^2, which when converted to equivalent 1 km, 25% duty cycle data, ranged from 0.64 to 0.98 mW/cm^2 with an average (from the 4 locations) of 0.82 mW/cm^2.

The Panel concluded that the measurement protocol and instrumentation used followed good engineering practice and that the variability in the measurement data (nearly 2 dB) was acceptable given the calibration procedures employed and differences in line of sight to the various measurement locations. Interestingly, in this same report (NRC 1979b), the panel made its own estimation of nominal and worst-case estimates of average power density on the main axis of a secondary sidelobe taking into account characteristics of the radar, and reported numbers of 8.2 μW/cm^2 (nominal) and 11 mW/cm^2 (worst-case) at 1 km, which are approximately an order of magnitude larger than the actual measurements but reassuring as conservative estimates. When factoring in more of the pulse-train specifics associated with enhanced search mode, the panel's estimate reduced to 1.8 μW/cm^2 at a distance of 1 km, which is only about 6 dB greater than the measurement data.

1979 NRC Exposure Level Report

This report (NRC 1979a) reviewed similar data measured by the Air Force in 1978 (August and October) at various points within, and up to 5 miles beyond, the restricted area, which is 1000 feet from the radar. It also considered analytical results for PAVE PAWS produced by the Environmental Protection Agency (EPA) in 1977. The August 1978 data were the same as reviewed in the 1979 NRC Engineering Panel report, although much more information was included in NRC 1979b. Specifically, at each of the 4 measurement sites, data from main-beam elevations of 3°, 6°, and 10° were evaluated along with both peak and average power-density recordings. The results showed a 344 μW/cm^2 peak power density at 3° elevation at the closest distance (1600 feet) dropping to a peak power density of 179 μW/cm² at 10° beam elevation with average power densities corresponding to 3.26 to 2.90 μW/cm², respectively, for these two beam elevations. At the largest distance (3900 feet), the 3° peak power density reduced to 48 μW/cm² with an average of 0.38 mW/cm², while at 6° beam elevation these numbers were 17 μW/cm² peak and 0.20 μW/cm² on average (the 10° beam elevation was not reported).

Measurements were also summarized for 21 sites in surrounding locations (Bourne, Sandwich, Mashpee, and Falmouth, MA ranging in distance from 1.0 to 22.2 km from the radar) on 2 days in October 1978 using both faces of the radar with an 18% duty cycle and 3° beam elevation. The committee compared these data to calculated values of electric field strength and average power density predicted to be emitted by the radar. It found that, in general, (1) exposures decreased with distance from the radar (but not uniformly), (2) measured values decreased more rapidly than calculated quantities farther from the source (likely due to attenuation not modeled from terrain, atmosphere, etc.), and (3) the ratio of peak to average power density varied significantly depending on beam elevation, azimuthal angle, and distance from the radar. More specifically, the committee concluded that average power density outside the Air Force base was not likely to exceed 1 mW/cm², the average power density at the exclusion fence surrounding PAVE PAWS was approximately 5 μW/cm², while the average measured intensity at locations where the public would most likely be exposed was 0.06 mW/cm² (Route 6, 1.0 km from the radar).

1839 EIG Engineering Report (#86-33)

In the fall (September 18-30) of 1986, time-averaged power-density measurements were recorded to document the RF exposures in the lighting and security camera areas within and around the security fence surrounding the PAVE PAWS radar facility on Cape Cod (EIG 1986). These data were requested because installation contracts for new lighting and camera systems at all four PAVE PAWS sites were being issued and construction workers would need access to locations directly in front of the radar. At the same time, the State of Massachusetts Department of Public Health was investigating possible causes of larger than normal cancer clusters on Cape Cod and the opportunity arose to also record RF power density levels at locations within the local community around the radar.

On site, measurements were reported for 17 lighting pole locations and 7 security camera positions immediately inside the security fence (approximately 120 feet from the radar face). Data were recorded at heights of 6 and 40 feet (above ground level) for the light poles and 6 and 20 feet (above ground level) for the security cameras. Additionally, ground-level (6 feet above) measurements were recorded at 10 other locations: 4 of these were outside the security fence but inside a secondary fence (between 120 and 200 feet from the radar face); 3 were outside the second fence (but within 250 feet of the radar); 2 were inside the security fence (one at the access gate lateral to the north face of the radar and the other directly in front of the north face 50-60 feet away); and 1 was not labeled on the map of measurement sites contained in the report but was listed in the data table of power-density recordings (presumably, it was associated with an 18th lighting pole not explicitly shown in the diagram, although data were only recorded at the 6 foot height). Power-density data were acquired at the 20 and 40 foot elevations (above the ground) using a bucket-loader truck that raised the test antenna and an operator to a height typically expected for maintenance personnel servicing either the security cameras (20 feet) or the lighting system (40 feet). A surveying transit was used to estimate the height of each measurement location relative to the bottom edge of the radar face. Because of the sloping terrain around the radar, these distances varied depending on the antenna test position. For the 6-foot (above ground) measurements, the heights ranged from even with to 14 feet below the bottom edge of the radar (at the greatest distance away, approximately 250 feet). The 20-foot elevations (at the security cameras) ranged from 5 to 14 feet above the bottom edge of the radar, while the 40-foot elevations (at the lighting poles) ranged from 31 to 45 feet above the radar's bottom edge.

The equipment used during the recording sessions was calibrated and certified by the Keesler Air Force Base Precision Measurement Equipment Laboratory based on specifications issued by the National Bureau of Standards. Results showed that the time-averaged power-density recordings had an overall accuracy of ± 2.0 dB when measuring a pulsed RF signal. When measurements were acquired, the PAVE PAWS radar was operating in its normal mode. The results showed that the highest time-averaged power-density at 6 feet above ground level was 1850 $\mu W/cm^2$ with a minimum of 38 $\mu W/cm^2$. The average was 529 $\mu W/cm^2$ at the 33 locations comprising the lighting poles, security cameras, and selected positions in and around the security fence. Data at the elevation of the 7 security camera locations (20 feet above ground) had a maximum power density of 745 $\mu W/cm^2$, a minimum of 100 $\mu W/cm^2$, and a mean of 273 $\mu W/cm^2$. Data at the elevation of the 17 lighting poles (40 feet above ground) had a maximum power density of 3716 $\mu W/cm^2$, a minimum of 127 $\mu W/cm^2$, and a mean of 1190 $\mu W/cm^2$.

The same equipment and procedures were used to measure power densities within surrounding Cape Cod communities at a height of 6 feet above the ground for all except two locations: (a) the Sandwich Fire Tower, which was 86 feet above the ground and in direct line of sight of the radar, and (b) the Otis Central Control Tower which was 106 feet above ground. Locations were selected by the Massachusetts Department of Public Health to represent areas of dense population and to augment data obtained from earlier measurement efforts (NRC 1979a, b). Once a location was selected, the test antenna was placed in direct line of sight of the radar whenever possible or in an open area to record maximum signal intensities. Data were measured at 15 locations in the surrounding communities ranging in distance from 1.2 to 8.8 miles from the radar. At 10 of the sites, power-density levels were below the measurable threshold of the equipment (determined to be 0.001 μW/cm^2). At locations where levels above this threshold were received, the largest value, 0.139 μW/cm^2, was documented at the Sandwich Fire Tower (3.2 miles away) at the 86-foot elevation, which was in direct line of site of the radar. The minimum measurable value, 0.003 μW/cm^2, was recorded at the Otis Central Tower (5.9 miles away elevated 106 feet above ground). Of the measurable levels at 5 sites, the average power density was 0.041 μW/cm^2. These locations averaged 3.2 miles away from the radar (ranging from 1.2 to 5.0 miles away).

2000 MITRE Report

In that report (MITRE 2000), two-dimensional peak and average power-density maps on the ground were presented for Cape Cod based on simulation studies of PAVE PAWS. Areas with line of sight to the radar were considered based on digital terrain elevation data for the Cape over a distance of 50 nautical miles. At these locations, the power density was calculated by considering the distance from the radar, the main beam elevation angle, and the time-averaging factors associated with the pulse-transmission scheme. For the peak-power calculation, a peak radiated power (543 kW), antenna gain (38.4 dB), and antenna pattern (beam at 3° elevation), along with a scan loss for off-broadside angles, were used to construct the maps. Generally, the peak-power densities in the 1-5 mile range from the radar cluster in the 1-10 μW/cm^2 level but fall off to 0.3-1.0 μW/cm^2 at distances 5-10 miles from the radar, although some variation with azimuthal angle is evident.

To estimate an average power-density map, the peak-power calculation was modified by the duty cycle, the scan revisiting time, and the azimuthal weighting factor. A duty cycle of 25% was used, a scan revisit factor (fraction of time the main beam is pointed in a particular direction given its beam width) of 0.066 was assumed, and an azimuthal-weighting factor that increased the average power 4 times (far from broadside) was applied. The map shows that beyond the first couple of miles from the radar, the average power density is within 0.01-0.1 mW/cm^2 and generally less than 1 mW/cm^2 within the first couple of miles (except very close to the radar).

To simplify the analysis, peak and average intensity maps that assume the terrain is flat at a height of either 0 or 100 feet above sea level (with the radar at a height of 270 feet above sea level) are also included in this report. These results show peak-power densities (assuming sea-level terrain) of 1.0-10.0 mW/cm^2 up to 5 miles or more away from the radar and a similar distribution of peak intensity at 100 feet above sea level with several isolated "hot spots" of 10-100 μW/cm^2 at distances up to 2 nautical miles from the radar. The average power-density maps at sea level yielded values between 0.01 and 0.1 μW/cm^2, with pockets having exposures between 0.1 and 0.3 μW/cm^2 approximately 2 miles from the radar. The 100-foot elevation map is similar but with more uniform exposure between 0.1 and 0.3 μW/cm^2 in a zone extending up to 4 miles from the radar with levels between 0.01 and 0.1 μW/cm^2 existing beyond this zone.

The Broadcast Signal Lab Report

During the first quarter of 2004, the Broadcast Signal Lab, Medfield, MA (BSL 2004a, b), executed a survey of RF emissions from the PAVE PAWS radar facility located at the Massachusetts Military Reservation on Cape Cod. The final test plan was approved by the PAVE PAWS Public Health Steering Group (PPPHSG) based on its solicitation for competitive proposals to complete the work. It consisted of three distinct tasks: (1) measurement of radar emissions during its normal operation at 50 open, publicly accessible locations throughout Cape Cod; (2) measurement of ambient emissions from all other sources in the VHF and UHF radiofrequency spectrum (30 MHz to 3 GHz) at 10 locations on Cape Cod (these data are discussed in the subsection on other environmental exposures); and (3) estimation of the radar exposure on Cape Cod based on calculations from a mathematical description of the PAVE PAWS antenna used in conjunction with propagation-modeling software provided by the MITRE Corporation.

All measurement methods and procedures were consistent with guidelines and consensus standards for performing such work, and the 50 field locations for assessing radar exposures were selected in consultation with the PPPHSG and the International Epidemiology Institute (IEI). A number of factors including (1) distance, (2) elevation, (3) population density, and (4) beam sweep coverage were considered in the site selection process. A detailed discussion of these parameters and the 50 field measurement locations is available elsewhere (BSL 2004a). Table 4-1, reproduced from the Final Test Plan (BSL 2004a), is a useful summary of the number of measurements fulfilling certain site criteria regarding height, intervening terrain, distance, etc. Maps of the specifics of locations by town, county, and approximate coordinates can be found in the test plan (BSL 2004a) and final report (BSL 2004b). The actual measurement site locations were modified slightly at the time of data acquisition to ensure safety, accommodate access issues, and improve measurement quality as determined by the field crew. Measurements were recorded for 90-minute periods (as six 15-minute routines where the detector was moved 3 feet after each acquisition sequence) in order to acquire realistic estimates of the average and peak radar emissions at each location.

Summary tables of power-density measurements at the 50 sites within the communities of Cape Cod show that the average values of the recordings at each location are extremely small. The largest average value was 0.035 μW/cm^2, which was recorded in Shawme Crowell State Park (site #15) at a location close to the

TABLE 4-1

Measurement Site Characteristic	No. of Sites	% of Sites
Measured with 30 ft-high antenna	15	30
Radio line of sight to radar	36	72
Terrain obstruction to radar	13	26
Man-made obstruction to radar	1	2
Beneath beam sweep	40	80
Beyond beam sweep (includes rear)	10	20
Less than 3 miles from radar	16	32
3 -10 miles away from radar	19	38
More than 10 miles away from radar	15	30
Barnstable County	44	88
Plymouth County	6	12

radar (less than 1 mile) that was relatively high (167 feet above sea level). The next highest sets of recording levels were nearly an order of magnitude lower (~0.002-0.005 $\mu W/cm^2$) and occurred at 4 other locations (site # 9, 21, 23, and 40) that generally tended to be closer to the radar (within 3 miles) although at variable heights (9-167 feet above sea level). One interesting exception was the Scargo Hill site whose average power density was 0.0038 $\mu W/cm^2$ at some 18.5 miles from the radar (65 foot elevation). Direct line from the radar to this location does cut across the coast line of the Cape, which may explain its relatively high (compared to radar distance) values, although the direct paths to sites 1-5 and 9 cut across mostly water (at distances ranging from 24 to 31 miles from the radar) and do not appear elevated compared to other locations. The 5 sites (# 9, 15, 21, 23, and 40) with the highest average values, not surprisingly, also shared the highest peak-power recordings in the 14-15 $\mu W/cm^2$ range. These peaks were, again, nearly an order of magnitude greater than peaks measured at the other 45 locations. The next cluster of peak values hovered in the range of 1.4-1.5 $\mu W/cm^2$ and were found at 10 other sites (# 14, 16, 18-20, 22, 25-27, 47), which were all within 6 miles of the radar and averaged a distance of 2.6 miles away.

Of the 5 sites with the highest average and peak power-density measurements, 4 were located in the sidelobe overlap zone (the 5th was in the northern face sweep zone). The majority of the 10 sites with the next highest levels of peak power density (all but one, #14) appeared in either the northern sweep (5 sites) or sidelobe overlap (4 sites) zones. Although there is an overall decrease in detected signal levels with distance, sites close to the radar (e.g., Route 6E Canal Overlook, #29, 1.9 miles away) can have low levels (0.135 $\mu W/cm^2$ peak), while locations at much further distances (e.g., Rock Harbor Parking, #5, 27.5 miles away) can exhibit relatively higher levels (0.153 $\mu W/cm^2$ peak), reflecting the importance of the intervening and local terrain in addition to the direct-line distance.

Comparisons between the current measurements and those recorded within the community in either 1978 (NRC 1979a,b) or 1986 (EIG 1986) show that the present equipment was significantly more sensitive and able to achieve recordings of weaker signals than was possible during the earlier efforts. At sites within 7 miles of the radar (previous studies did not measure responses at greater distances), the 2004 data are consistently lower than the prior recordings. Specifically, the average power-density readings for sites within 7 miles were below 0.01 $\mu W/cm^2$ in the 2004 data with many points below 0.0001 $\mu W/cm^2$. The 1978 and 1986 results showed average power densities between 0.1 and 0.01 $\mu W/cm^2$ for most sites within 3 miles of the radar with the largest value being above 0.1 $\mu W/cm^2$ at 3-plus miles from the radar in the 1986 data set. About one-half (4 locations) of the 1978 readings were between 0.01 and 0.001 $\mu W/cm^2$ and a third (2 sites) of the 1986 results were in this range (levels below 0.001 $\mu W/cm^2$ were not considered to be measurable with the instrumentation used in the earlier power-density surveys).

The modeling effort produced results from the propagation model combined with an antenna pattern to generate a matrix of public-exposure estimates to the PAVE PAWS emissions. The propagation model was evaluated by comparing its predictions to measurements obtained in 3849 latitude-longitude cells acquired during a 250-mile drive test. The comparison between computed and measured values indicates that the model is accurate to within the uncertainty of the measurements. Specifically, the average variance between the propagation model prediction and the drive-test recordings was –1.6 dB. A MITRE antenna model was used to create an antenna pattern for the propagation software. This led to the creation of the exposure matrix for Cape Cod, indexed by latitude and longitude, containing values for each position in terms of $dB\mu W/cm^2$. To validate the results when combining the antenna and propagation models, measurements at 35 sites were compared with the corresponding calculations. Departure between the model estimates and the measured data was 6.5 dB on average over all sites with a standard deviation of 8.5 dB. Excluding 4 sites behind the radar reduced the average departure to 5.9 dB. Overall, the model estimates slightly over-predicted the levels measured in the field.

COPPS Data

The Coalition to Operate PAVE PAWS Safely (COPPS) performed power density measurements on Cape Cod at 26 sites (23 firehouses, 2 commuter parking lots, and 1 lighthouse) spanning 311 days beginning in May of 2003. Analysis of existing data as of April 2004 was made available to the committee through website postings (see COOPS 2004). Specifically, a town-by-town map of measurement sites, a color-coded rank ordering by measured (http://www.pavepaws.com/measured_sites.pdf) peak-power density (in $\mu W/cm^2$) of the towns where data were recorded (http://www.pavepaws.com/peak_measurements.pdf), and plots of peak-power versus various parameters of interest (e.g. elevation, distance, obstruction, bearing, weather, season) (http://www.pavepaws.com/additional_findings.pdf) were provided. The results are generally consistent with the Broadcast Signal Lab (BSL) report (BSL 2004b) in a number of important respects, although the details of specific measurements may be different (and are not easily compared given information presently available). The COPPS data show peak-power densities of less than 1 μW/cm^2 regardless of location, with more than 75% of the 40 peak values reported (http://www.pavepaws.com/additional_findings.pdf) being near or below 0.1 μW/cm^2. These levels are lower than the highest peak values recorded by BSL but consistent with the second and third tiers of peak values into which the majority of the BSL measurement sites also fall. Both the BSL and the COPPS data sets reveal that the locations with higher peak power-density recordings are preferentially distributed in the northern sector of the radar. Both data sets also indicate that some of the highest peak power-densities occur at locations having the greater distances from the radar, especially when the straight-line distance from the radar to the measurement site cuts across the coastal waters of Cape Cod Bay. Other findings in the COPPS data are consistent with expectations (and the BSL results) that the degree of topological obstruction between the measurement location and the radar facility is a critical determinant of the measured power-density exposure. Positive correlations between power-density levels and clear weather and the fall/winter seasons were reported in the COPPS analysis. Again, these results are reasonably anticipated from the perspective that inclement weather, increased foliage, and other factors are expected forms of obstruction to radiowave propagation at PAVE PAWS frequencies.

POPULATION EXPOSURE TO OTHER RF SOURCES

It is informative to consider general and specific population exposures to other RF sources that have some similarities to PAVE PAWS for gaining a perspective on the power-density levels described in this chapter. Interestingly, the 1979 engineering report (NRC 1979b) compared the transmitter characteristics of PAVE PAWS with two other radar installations that radiate much higher peak powers (megawatts) and higher (for one) and lower (for the other) average powers but no power-density measurements on the ground were reported. In NRC (1979a), data obtained by the EPA in the 1970s (EPA 1978a) from population exposures in 12 large U.S. cities over the FM and TV broadcast bands (54-900 MHz) were summarized. These studies estimated that the median exposure in urban areas was 0.005 μW/cm^2 and that 95% of the urban population is exposed to less than 0.1 mW/cm^2. These numbers are generally consistent with the PAVE PAWS data in terms of power-density exposure. The EPA has also reported power-density measurements in tall buildings that either house, or are near, broadcast antennas (EPA 1978b). These studies show that, depending on the location within the building and/or the degree of shielding from the antenna site, values can range from less than 1 μW/cm^2 to 97 μW/cm^2 on the inside to as much as 230 μW/cm^2 outside (and near) the source.

The Broadcast Signal Laboratory study (BSL 2004b) selected 10 sites for assessing the ambient RF exposure that were chosen in consultation with the PPPHSG and IEI, and a detailed discussion of the rationale for their selection appears in BSL 2004a. A brackcting strategy was adopted because, as with the PAVE PAWS radar, itself, the levels of ambient VHF and UHF radiation measured at a given location depend on distance from the source, the intervening terrain, and local topological environment. At least a dozen FM radio stations (a number of which operate with radiated powers of 45 kW or more) and a couple of TV stations (including one UHF station licensed to operate at 1150 kW) are present in the area. Numerous (more than 5000) land-mobile facilities, typically licensed for 100 to 1000 watts per channel emissions (e.g. police, fire, business systems) were also identified. The final 10 sites selected attempted to represent high, low, and middle-level exposure conditions to these sources as a means of providing a representative sampling of the ambient RF exposures on Cape Cod. The propagation model was used to define the power levels and associated locations that would be expected to bracket exposure conditions given the relevant sources of radiation identified. Descriptions of the final site selections are reported (BSL 2004b). Four pre-measurement bracketing level estimates spanning 10^{-1}-10^{-5} (i.e., 10^{-1}, 10^{-2}, 10^{-4}, 10^{-5}) μW/cm^2 were used to categorize the 10 sites.

Measurements at the 10 sites selected to bracket the levels of ambient RF exposure across the VHF and UHF signal bands were reported in terms of the maximum permissible exposure (MPE) for low (30-320 MHz), mid (320-950 MHz), and high (950-3050 MHz) bands that were combined into a final figure of merit that could be compared to the MPE for PAVE PAWS. These findings were tabulated as dB below MPE. For example, the MPE for PAVE PAWS (determined by applying the IEEE weighting for the 420-450 MHz band) is 290 μW/cm^2 thus the maximum and minimum measured power densities from PAVE PAWS (at the 50 measurement sites) convert to –48 dB (below MPE) and –103 dB, respectively, whereas the equivalent values for the ambient signals from the 10 sites resulted in maximum and minimum measured values of –19 dB and –50 dB (relative to MPE). These results indicate that the ambient exposure level at the most energetic site (near the FM tower at the Exit 6 commuter parking lot on Route 6) was 100 times less that the MPE for the general population. The other sites were an additional 10 to 1000 times less exposed to VHF and UHF emissions. Overall, the highest average PAVE PAWS exposure at any of the 50 measurement sites was comparable to the lowest ambient levels observed at the 10 locations sampled. Further, the variation between ambient sites was no more than a factor of 1000 (30 dB), while the lowest observed PAVE PAWS signal was about a million (60 dB) times weaker than the highest.

Cell-phone exposure has generated considerable interest, and there have been a number of efforts to quantify exposure. Two issues are relevant: first, the exposure level to the cell-phone user due to the radiated power during a call, and second, the exposure of the general population to cell-phone relay stations. The former issue has been studied in some depth. The power radiated for a typical cell phone is generally less that 0.6 W and it has been shown that approximately half of this power is absorbed in the hand and head of the user, leading to specific absorption rates on the order of 1 W/kg (Jensen and Rahmat-Samii 1995).

The committee is not aware of extensive studies of power-density levels near relay stations (base stations) in the United States, but there are 3 informative evaluations from Canada, the United Kingdom, and Australia.[1] In Vancouver, Canada, Thansandote and others (1999) measured RF levels in 3 schools with base stations on or near their campuses and 2 schools with no antennas nearby that served as controls. Maximum measured exposures in the 3 schools with adjacency ranged from 0.16 μW/cm^2 to 2.6 μW/cm^2 depending on the location of the station (2.6 mW/cm^2 resulted from a base station on the roof of a school). The 2 schools without base stations in their vicinity had a maximum power density of 0.01 μW/cm^2.

The U.K. National Radiological Protection Board measured RF power-density levels at 118 sites around 17 base stations in 2000 (NRPB 2000). Typical levels were less than 0.1 μW/cm^2, while the highest recorded power density at any location was 0.83 μW/cm^2. The measured upper bound on power density as a function of distance from the base of a building or tower with an antenna showed an increase (from just below 0.01 μW/cm^2) for approximately the first 60 meters (to a level approaching 0.1 μW/cm^2) followed by a more gradual decrease (back to 0.01 μW/cm^2) up to 200 meters away. The power-density auditing program in the U.K. has continued with data available from recording in 2001-2004. Results are summarized in terms of a percentage of the ICNIRP (International Commission for Non-ionizing Radiation Protection) power-density guidelines for a band exposure equivalent that factors in the changing safety standard as a function of frequency from 400 MHz to 2 GHz, which for the PAVE PAWS emissions would represent an upper limit of 200 mW/cm^2 (at 400 MHz). The highest values reported at any location for each year would represent PAVE PAWS-like exposures (i.e., band equivalent exposure extrapolated to 400 MHz) of 0.72 μW/cm^2, 0.27 μW/cm^2, 0.32 μW/cm^2, and 0.26 μW/cm^2 for the years 2001, 2002, 2003, and 2004, respectively.

The survey of GSM base stations by the Australian Radiation Protection and Nuclear Safety Agency measured 13 sites that yielded an average power density

---

[1]From the Website http://www.mcw.edu/gcrc/cop/cell-phone-health-FAQ/toc.html and http://www.ofcom.org.uk/consumer_guides/mob_phone_base_stat. Information available 10/25/04.

of less than 0.1 μW/cm^2; the highest power density measured was less than 0.2 cm^2.

SUMMARY OF PAVE PAWS EXPOSURE DATA

While not extensive per se, the power-density measurements that do exist for the PAVE PAWS radar on Cape Cod, which have been recorded by different groups on a number of separate occasions over the radar's operational life, are generally consistent with expectations. The most extensive efforts in terms of the number and distribution of measurement sites are the recently issued 2004 reports (BSL 2004a, b; COPPS 2004), which have focused on assessments of exposure levels within the surrounding communities on Cape Cod. These studies reveal heterogeneity in the measured levels that demonstrate the strong influence of the intervening terrain and local topological environment on the recorded power densities and consequently the public exposure. While there is a general decrease in the power-density levels measured with distance from the radar that is evident in the various datasets that have been assembled, it is also clear that locations much farther (10-plus miles) from the radar can be exposed to higher levels of radiation than sites nearer (within 3 miles) the facility depending on topographical features. Certainly, both the BSL and the COPPS data (BSL 2004a, b; COPPS 2004) demonstrate this characteristic at selected locations. Nonetheless, the average power densities have been consistently below 0.1 mW/cm^2 and generally in the 0.001-0.01 mW/cm^2 range at locations where the public would be expected to be exposed. The simulation studies that have been reported estimate higher values overall, but for the most part still suggest that the exposure of the public is less than 1 μW/cm^2 in terms of average power density. Measured peak levels are generally less than 1 μW/cm^2, although values as high as 15 μW/cm^2 have been found at a few elevated locations near the radar. These levels are well below the widely accepted thresholds for the power absorption that would be required to generate harmful thermal effects in tissue. The degree to which they may cause other biological effects in cells and animals is discussed in Chapters 6 and 7, respectively, and the implications for human health are developed in Chapters 8 and 9. The degree to which emissions from PAVE PAWS may contain special characteristics pertaining to its time-domain waveform is described in the following chapter (Chapter 5).

REFERENCES

ARPANSA (Australian Radiation Protection and Nuclear Safety Agency). 2000. Line, P., W.A. Cornelius, M.J. Bangay and M. Grollo. Levels of Radiofrequency Radiation from GSM Mobile Telephone Base Stations (Tech Rep 129). Available from http://www.arpansa.gov.au/pubs/eme_comitee/rfrep129.pdf as of 10/25/04.

BSL (Broadcast Signal Lab). 2004a. Survey of RF Energy Field Emissions from the PAVE PAWS Radar Located at Cape Cod Air Force Station, Massachusetts: Final Test Plan, prepared in consultation with International Epidemiology Institute, Rockville, MD, by Broadcast Signal Lab, Medfield MA.

BSL. 2004b. A Survey of Radio Frequency Energy Field Emissions from the Cape Cod Air Force Station PAVE PAWS Radar Facility: Final Test Report, prepared for PAVE PAWS Public Health Steering Group by Broadcast Signal Lab, Medfield MA.

COPPS (Coalition to Operate PAVE PAWS Safely). 2004. Available from http://www.pavepaws.com/measured_sites.pdf; http://www.pavepaws.com/peak_measurements.pdf; http://www.pavepaws.com/additional_findings.pdf.

EIG (Engineering Installation Group). 1986. Radio Frequency Radiation Survey for the AN/FPS-115 PAVE PAWS Radar Cape Cod MA, 18-30 Sep 86, Report #86-33, 1839 Engineering Installation Group, Keesler Air Force Base, Biloxi, MS.

EPA (Environmental Protection Agency). 1978a. Population Exposure to VHF and UHF Broadcast Radiation in the United States, R.A. Tell and E.D. Mantiply, EPA Technical Report ORP/EAD 78-5.

EPA. 1978b. Measurements of Radiofrequency Field Intensities in Buildings with Close Proximity to Broadcast Stations, R.A. Tell and N.N. Hankin, EPA Technical Report ORP/EAD 78-3, March 1978.

Jensen, M. and Y. Rahmat-Samii. 1995. Interaction of handset antennas and a human in personal communications. P IEEE 83:7-17.

MITRE. 2000. RF Power Density Exposure at Ground Level for the PAVE PAWS Radar at Cape Cod—Questions and Answers. MITRE Technical Report, MITRE Corporation, Bedford, MA.

NRC (National Research Council). 1979a. Analysis of the Exposure Levels and Potential Biologic Effects of the PAVE PAWS Radar System. Washington, DC: National Academy Press.

NRC. 1979b. Radiation Intensity of the PAVE PAWS Radar System, Engineering Panel on the PAVE PAWS Radar System, National Research Council, Final Report. Washington, DC: National Academy Press.

NRPB (National Radiological Protection Board). June 2000. Mann, S.M., T.G. Cooper, S.G. Allen, R.P. Blackwell, and A.J. Lowe. Exposure to Radio Waves near Mobile Phone Base Stations. Chilton, U.K.: National Radiological Protection Board.

Thansandote, A., G.B. Gajda, and D.W. Lecuyer. 1999. Radiofrequency radiation in five Vancouver schools: exposure standards not exceeded. Can Med Assoc J 160:1311-1312.

5

PAVE PAWS Exposure Conditions

EXPOSURE CHARACTERISTICS

The Radar and Its Signal

To briefly review the radar description provided in Chapter 4, the PAVE PAWS radar is pulsed; it transmits a pulse sequence, and then is quiescent while the radar echoes return. Then another pulse sequence is transmitted and the process is repeated. Pulse widths vary from 250 microseconds to 16 milliseconds, and the listening period is at least 38 milliseconds. Some pulses vary in carrier frequency during the pulse, using a technique called "chirping" to improve range resolution. The carrier frequency varies from 420 MHz to 450 MHz. Signal bandwidths at the input to the final amplifiers range from 8 kHz for the narrowest pulse to 125 Hz for the widest pulse. The "chirp" bandwidth is 2 MHz.

The antenna is comprised of two phased arrays, each located on a planar face, tilted back 20 degrees from vertical. The elements are located on a regular hexagonal lattice, and there are 1792 active elements per face and 885 passive elements. The elements are deployed in quadrantal symmetry to allow precise monopulse tracking of targets. The diameter of each array is approximately 72.5 ft. Each array face will support many more elements, but that expansion has not occurred, and the committee has been informed that there are no plans to expand the number of elements on the array. The elements are bent dipoles, each supported by a two-post balun. Bending down the arms avoids blind angles produced by dipole-balun mode interference. Crossed dipoles are used, thus providing both vertical and horizontal polarization. Both arms are hot; the two conductors supporting the dipole arms constitute a balun, which transforms the balanced dipole arm connection to an unbalanced coax; the latter connects to the module contain-

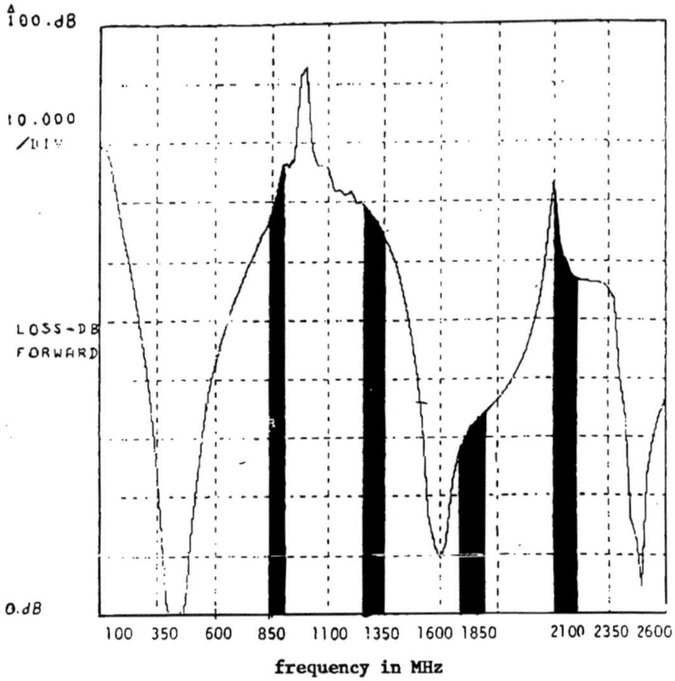

FIGURE 5-1 Measured module filter response. The filter minimizes the interference of PAVE PAWS to services such as cell phones, GPS, and other communications systems. Filter bandwidth is 115 MHz at 10 dB points. This figure shows the pass band (low attenuation for the radar frequencies [425-450 MHz], and high attenuation for harmonics; attenuation of 70 dB for second harmonic at 850 MHz; 68 dB for third harmonic at 1350 MHz; 31 dB for 4th harmonic at 1800 MHz; and 59 dB for fifth harmonic at 2100 MHz). This filter insures that the PAVE PAWS signal has very small harmonic content. Figure is reproduced with permission from Raytheon Company.

ing the amplifier. This is a well known and widely used technology. Each quadrant has 14 subarrays, and each subarray has 32 transmit-receive (TR) modules and 32 active elements. The array main beam is steered by applying phase shift to each TR module. All modules are excited simultaneously; beam scan is provided by phase shift. Sidelobe levels are held 20 dB below the main beam by tapering: the inner subarrays have all active elements, while middle subarrays have some passive elements, and outer subarrays have more passive elements. Thus an amplitude taper is produced over the array (see Figure 5-4). Each TR module includes a bandpass filter to minimize interference caused by the radar. Filter bandwidth is 115 MHz at 10 dB points; the measured filter response is shown in Figure 5-1. Note that the filter skirts around 420 and 450 MHz are quite steep. This filter affects the pulse buildup and decay.

Measurements of the exciter output RF waveform are shown in Figures 5-2a, 5-2b, and 5-2c. Note that these appear inverted, due to the measuring equipment using negative detection. In Figure 5-2b the rise time is approximately 1 microsecond, with a gradual rise and gradual leveling off. Finally Figure 5-2c shows fall time, which is much slower, roughly 12 microseconds, and therefore no abrupt changes are evident. Note that the rise-time bandwidth is narrow (roughly 2 MHz).

WAVEFORM DECAY

Each PAVE PAWS dual polarized element is driven by a Class C (highly non-linear) amplifier; a second Class C amplifier drives all 64 element amplifiers in a sub-array; a third Class C amplifier drives all 56 sub-arrays. All of these amplifiers are "on" all the time; the radar exciter unit applies the waveform (pulse) to the amplifier driving the sub-arrays. Beam steering is provided by a phaser at each element module. Class C amplifiers tend to sharpen up transients, so the exciter buildup and decay times of roughly 1 and 12 microsec (Figures 5.2b and 5.2c) become typically 200 and 1000 nsec for the radiated signal.

An examination of Phase IV buildup and decay pairs shows that these times are roughly equal, about 200 nsec (see Figures 5-6 and 5-11, which are expansions of Figures 3-48e and 3-53e in the AFRL (2003) Phase IV Time-Domain Waveform Characterization report. In a number of cases, a trailing-edge spike is larger than the main waveform. All of these are at wide angles from the main beam, where the sidelobe envelope is much lower. These spikes are probably due to two factors working together. First, the individual Class C amplifiers do not turn off alike, due to different resonant circuit Q (ratio of stored energy to dissipated energy), different gain, and other factors. Second, when all element contributions add in phase, the main beam is produced. Sidelobes and nulls are produced when the element contributions partly cancel, producing sidelobe levels away from the main beam 30-40 dB down (in power down to 0.001-0.0001 from 1). If amplifiers have typical gain variation of ±1 dB or even more, field-strength levels well above the steady-state sidelobe level are possible as the elements variously shut down. This is substantiated by calculations made by Tomlin for the PPPHSG.[1] It should be noted that the discrete delay is not a possible cause here.

Signal Propagation

The radar signal power decays as $1/R^2$ in free space, but multipath signals are usually a problem at UHF. Multipath is caused by diffraction and reflection. For example, if a hill or slope blocks the direct radar signal, the signal can be re-

[1]Jim Tomlin, Charts and Notes prepared for the PPPHSG, January 6, 2004. Jim Tomlin is a technical advisor to the PPPHSG.

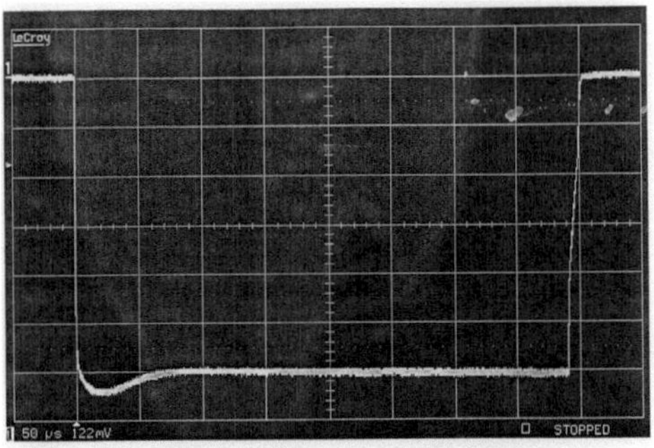

FIGURE 5-2a Entire exciter waveform pulse (negative direction) taken from RF monitor example in T.O. 31P6-2FPS115-51, Chapter 6.

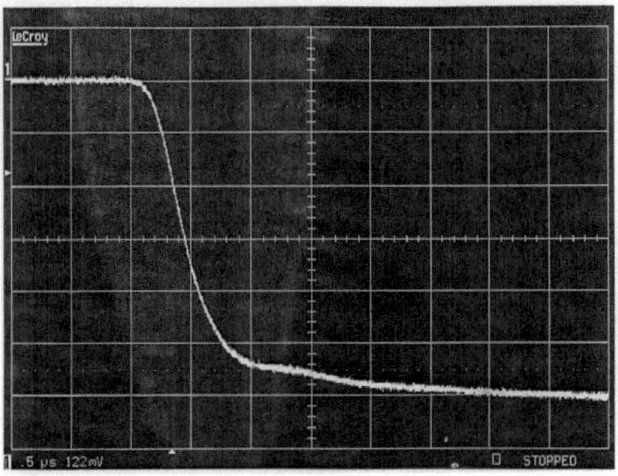

FIGURE 5-2b Exciter output pulse (negative direction) rise time taken from RF monitor example in T.O. 31P6-2FPS115-51, Chapter 6.

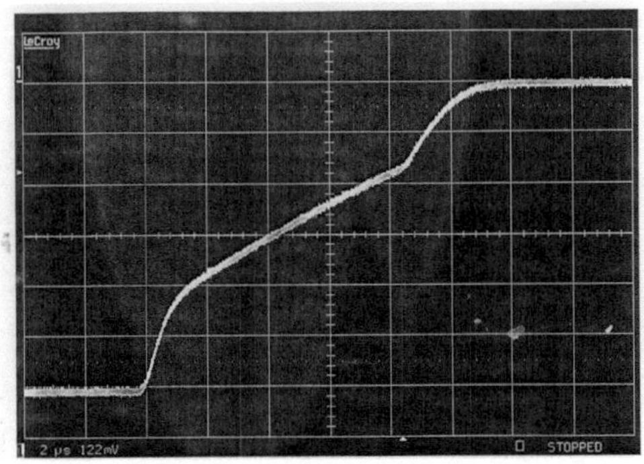

FIGURE 5-2c Exciter output pulse fall time (negative direction) taken from RF monitor example in T.O. 31P6-2FPS115-51, Chapter 6.

flected or diffracted by towers, buildings, power lines, or other obstacles and can reach areas that are not line-of-sight from the radar. These multipath signals need not be in the direction of the observer; they may arrive sideways. Reflected multipath is provided by many types of metallic structures: metal lighting poles; fence poles and wire; rebars in concrete; and telephone cable and power wires, to mention a few. These multipath signals are well known, and can easily disrupt the $1/R^2$ energy falloff. Almost certainly the anomalous Phase IV measurements are due to multipath.

The question of surface waves has been raised by concerned Cape Cod citizens. However, it has been accepted for 50 years that the longitudinal field component that may exist at HF, VHF, and UHF is not a surface wave. Sommerfeld in his classic 1909 work (Sommerfeld 1914) made a sign error, according to Norton (1937) and others in several papers. There is recent evidence by Collin (2004) that the original sign was correct, but in either case the earth does not support a surface wave. The waves are now called "ground waves"; the vertical electric field tilts just enough to satisfy the lossy boundary conditions in the ground. The term "surface wave" has for at least 40 years referred to slow waves, waves with velocity less than that of light, and usually supported by reactive surfaces. At 435 MHz, ground waves attenuate rapidly. Using sandy soil (epsilon = 10, σ = .002 S/m), the radiation is attenuated 40 dB (0.0001 in power, 0.01 in field) over a distance of 120 m (400 ft) . Outside the safety fence, the ground wave should be too small to measure. Use of the formulas of Baños, as suggested, is not recommended. They are excellent for dipoles in earth, or dipoles very close to earth, but are error prone for separations of a wavelength or more.

Although it has been implied that the measured radial fields (radial waves) represent a new and unexpected phenomenon, they are just the result of multipath and simple geometry. As sketched in Figure 5-3a, the line of sight during the Phase IV measurements was not horizontal, but was inclined down about 2 degrees. Since the electric field at the measurement point is perpendicular to the line of sight, it can be resolved into a large vertical component and a small horizontal or longitudinal component. For many points on Cape Cod, the line of sight angle is larger, leading to a larger longitudinal component. At UHF frequencies multipath is a well known and serious problem. Metallic structures including towers, power lines, and rebars in concrete, that are inclined or curved, will convert vertically polarized fields into horizontally polarized fields. Even curved earth hills can produce polarization conversion. It is difficult if not impossible to compute the combined multipath effects in a given environment. The multipath contributions and the horizontal field due to line of sight tilt combine to produce a longitudinal E field. (See Figure 5-3b taken from 3-92c, and Figure 5-3c taken from 3-96c of the Phase IV report.) Because this field is parallel to ground, and the sensor is roughly two wavelengths above it, there are also ground reflection effects. The derivative sensors used in the measurements have very broad pat-

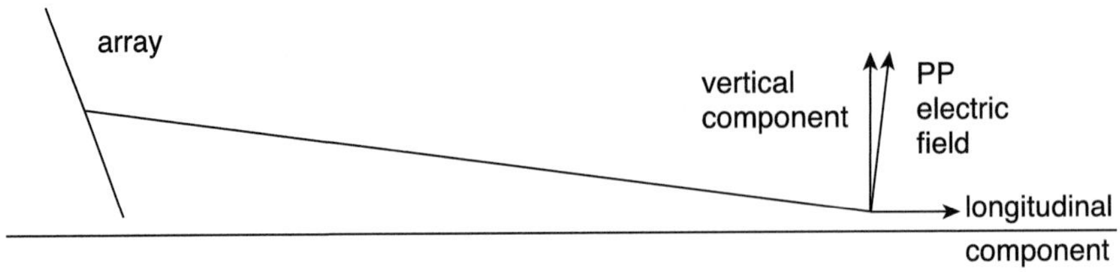

FIGURE 5-3a Vector resolution of PP electric field into vertical and longitudinal component. NOTE: This figure has been changed since the original prepublication version to correct an error.

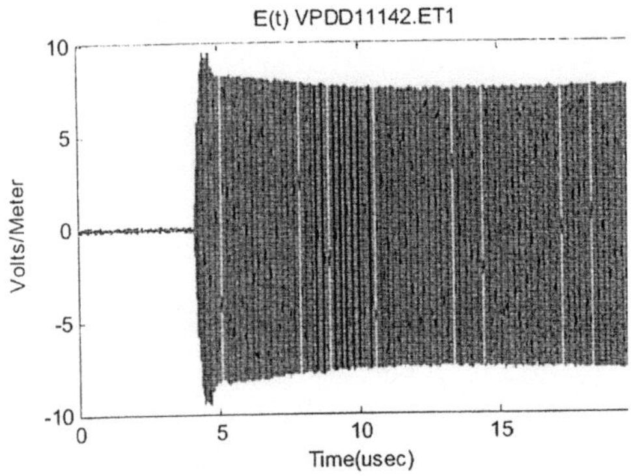

FIGURE 5-3b Vertical field amplitude (from 3-92.c of the waveform report).

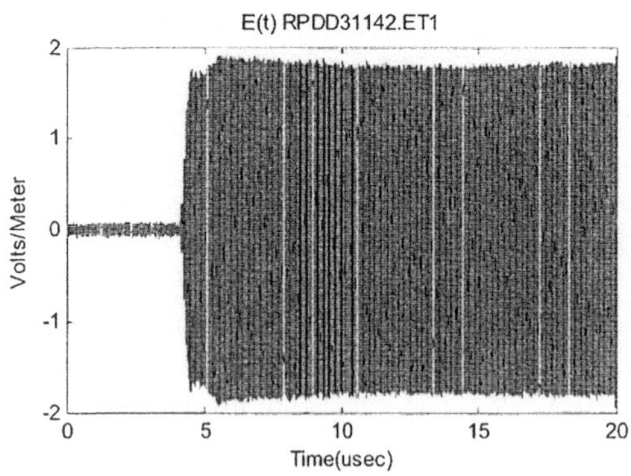

FIGURE 5-3c Radial field amplitude (from 3-96.c of the waveform report).

terns, and pickup signals from many directions. It would be better physics to call the field a "longitudinal component."

Time Delay in Large Antennas

It was suggested by Dr. Albanese[2] that phased array radars are different from other radars due to their discrete time delay. This is one of the key issues relating to PAVE PAWS effects and it is examined thoroughly here. In the far-field of a large phased array or reflector antenna, the contributions from all parts of the antenna arrive at the main beam peak simultaneously (broadside beam for a phased array). At other angles, there is a differential delay as the energy from the closest part of the array or reflector arrives first, while that from the farthest part arrives last. Of course this applies primarily to pulsed signals. For the reflector, the pattern buildup starts at 90 degrees from the axis due to the currents on the reflector edge (see Hansen 1987; Hansen and Kramer 1992) and the pattern builds up slowly and continuously. With a phased array, the pattern builds up slowly but in a discrete stair-step fashion. Due to the large number of elements, the stair has many very small steps. As expected, the effect of those small steps does not appear in the waveform measurement (see Figure 5-6).

In the near-field region (distance generally less than 2000 ft), there are delays also in forming the main beam. For the phased array, the near-field stair-step buildup of the waveform is expected to be very close to a continuous buildup. For a reflector antenna, the near-field buildup is more complex.

At angles away from the normal to the face of the array, there is a time delay between energy arriving from the closest element and that arriving from the farthest element, since all elements are turned on simultaneously. This delay, in amplitude and phase, has been computed for the PAVE PAWS radar. The active array is 72.5 feet wide. A worst case scenario is postulated: the beam is scanned to the farthest left position of 60 degrees azimuth, and to the lowest elevation position –17 degrees (3 degrees above the horizon). The observation point is 60 degrees right azimuth, and –20 degrees elevation (on the horizon). A computer program was written to calculate the waveform buildup. The x-y coordinates of the 1792 excited elements were provided on a disk by Mitre. Figure 5-4 shows the positions of the active elements. They are closely spaced around the center but space tapering is used in the outer portion, to control the sidelobe level.

---

[2] Dr. Albanese presented his theories to the committee in open sessions of committee meetings on several occasions and by letter. A number of causal hypotheses were also presented in a letter to the PPPHSG dated July 30, 2001, and the committee has also evaluated responses to these hypotheses made by Dr. Robert Adair in a journal commentary (Adair, R. K. 2003. Environmental Objections to the PAVE PAWS Radar System: A Scientific Review. Radiat Res 159:128-134).

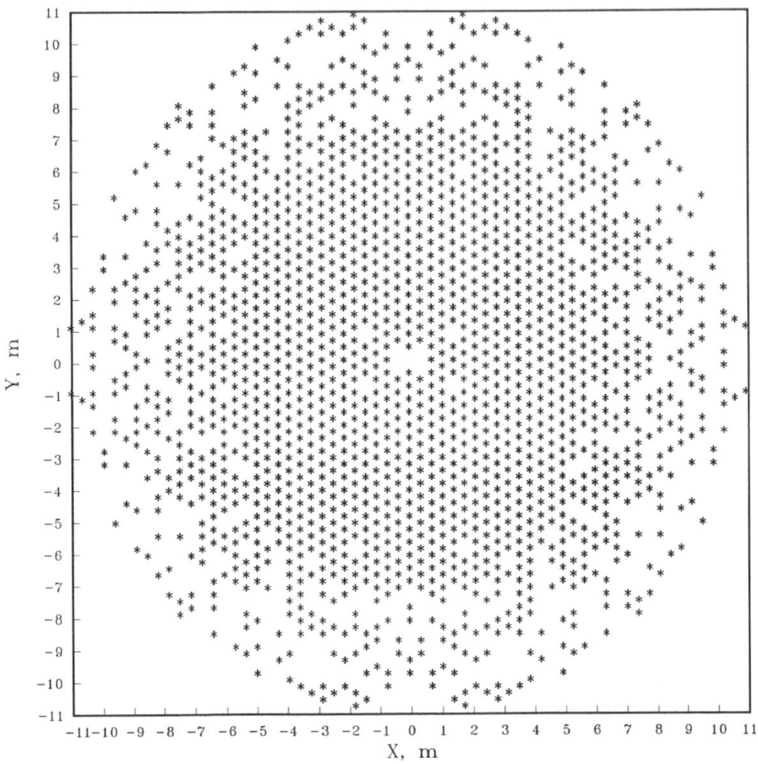

FIGURE 5-4 Positions of the active PAVE PAWS elements.

Using the observation direction, the projected distance from each element to the first element (center row, right side) is calculated for the 1792 elements. The projected distances are calculated using direction cosines, where the polar angle θ is measured from the normal to the array face, and the azimuth angle Φ is measured in a horizontal plane from normal to the surface. Since those with the smallest projected separation contribute first, the array of projected separations is ordered in ascending order by a well-known subroutine called HEAPSORT. Because the excitation phases associated with the main beam position affect the phase of the buildup, the x and y coordinates are sorted in consonance with the projected separations. The delay time across the array is roughly 74 nanoseconds; the delay for the worst case mentioned above is 60 nanoseconds. In order to get a fine-grain picture of how the elements contribute, the time line is divided into 12,000 steps of roughly 5 picoseconds each. The time stepping starts at 0 at the right-most reference element and proceeds on. When it has encompassed the next closest element, the contribution of that element with the appropriate scan phase and unity amplitude is added in. As the time step proceeds, more and more elements are added until the opposite side of the array is reached. Since the phase and amplitude are both a result of stair-step contributions, no smoothing or interpolation has been used in the plots. Figure 5-5a shows the amplitude buildup for the extreme case mentioned above.

This curve is made of 1790 stair steps, although they are not visible due to the scale. But a portion of the delay time is expanded in Figure 5-5b, where the steps are clearly visible.

The delays range from 5 psec (or less, as this is the least count) to roughly 100 psec. Note that the height of the large steps is roughly 4×10^{-3} times the final height, and that is the level of a far-out sidelobe, perhaps 1×10^{-4} times the main beam power. So a typical step represents a power step of the order of 4×10^{-7} times the main beam power. Figure 5-5c gives the phase buildup; the first step is an artifact due to the plotting program starting at zero phase. Note that the calculations were made along one of the four directions (axes) of symmetry, which is a worst case. If a slightly different azimuth angle had been used, the distance steps would be less regular, and the buildup would be even smoother. These incremental delays, as radiation from antenna elements appear, are very much smaller by comparison to the rise times (less than a factor of a thousand) produced by the TR modules, as discussed below.

Total phase change during buildup is very modest. Thus the phased-array delay is expected to produce a negligible effect; the waveform rise time is very gradual. Buildup is similar to that of a large parabolic dish antenna system.

The PAVE PAWS Phase IV measurement program produced data that are relevant to the effects of time delay, both in arrays, and in reflectors. All of the graphs presented below are excerpts of the digital data provided to the committee by AFRL. First, the PP waveform buildup at 60 degrees from the array normal (from Figure 3-48e of the waveform report) (AFRL 2003) is shown in Figure 5-6.

The PP waveform buildup near normal (from Figure 3-46b in the waveform report) is shown in Figure 5-7. At center frequency of 435 MHz the maximum delay in Figure 5-6 is 63.8 nsec, or 28 carrier cycles. The time-delay period extends over roughly one-third of the rise time in Figure 5-6. As expected from the buildup calculations just discussed, there is no apparent indication in Figure 5-6 that the discrete delay has any effect. Figure 5-7, near normal, incurs negligible delay; the slight irregularities in amplitude occur equally in both. Before buildup starts, the data show noise and interference; this probably accounts for the later small irregularities.

The committee notes that the Air Force Phase IV report revisions incorrectly assume time-delay beam steering, instead of phase beam steering.

For comparison Figure 5-8 shows buildup for a single PP element. The envelope is smoother, but the pre-buildup noise is smaller than in Figures 5-6 and 5-7. Similarly, the Phase IV measurements on two PP elements, with an introduced 50 nsec delay, are indistinguishable from those of a single element.

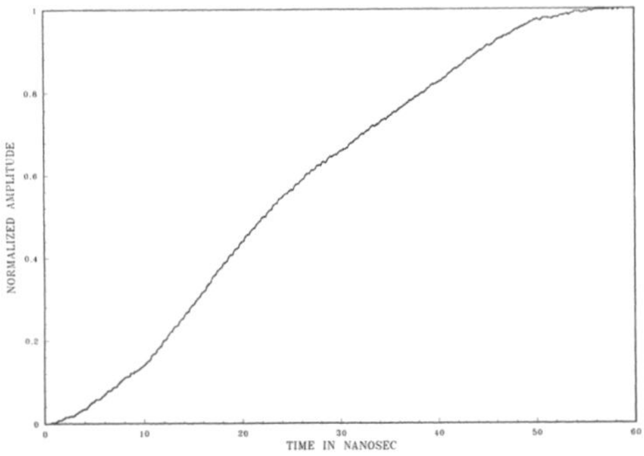

FIGURE 5-5a PAVE PAWS waveform buildup, az = 60 degrees, el = 0 degrees.

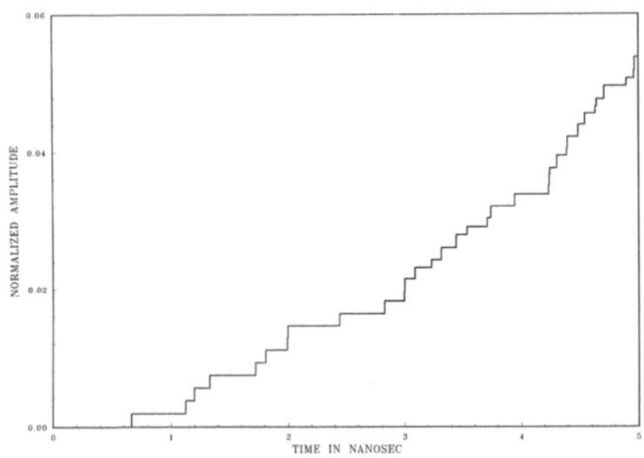

FIGURE 5-5b PAVE PAWS waveform buildup, expanded scale.

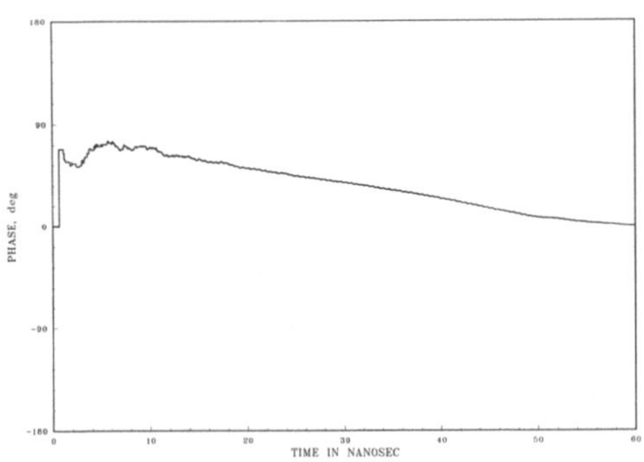

FIGURE 5-5c PAVE PAWS waveform buildup. az = 60 degrees, el = 0 degrees.

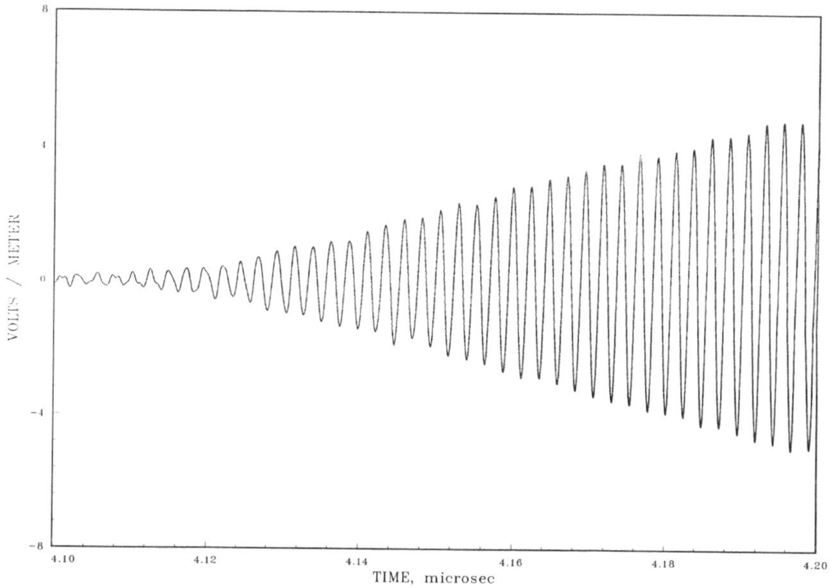

FIGURE 5-6 Array buildup; 60 degrees off normal.

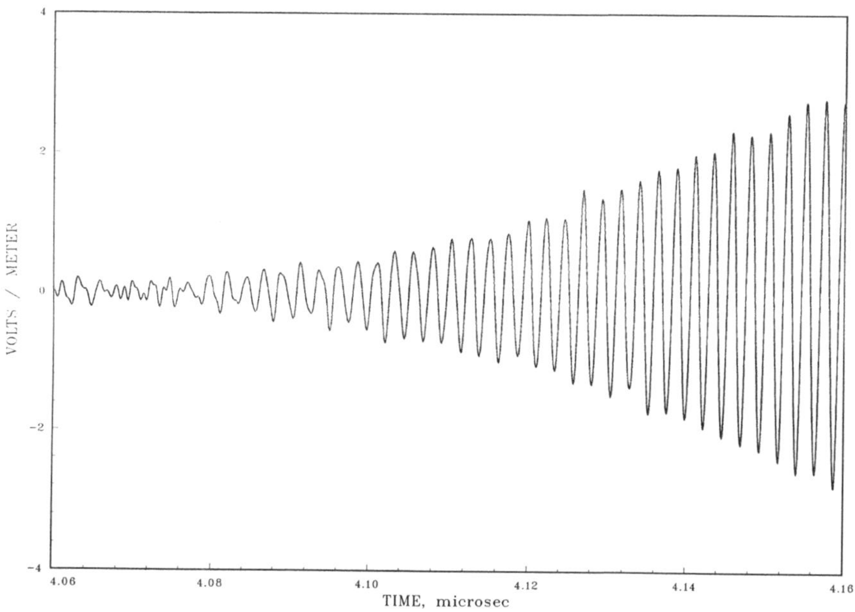

FIGURE 5-7 Array buildup; near normal to array face.

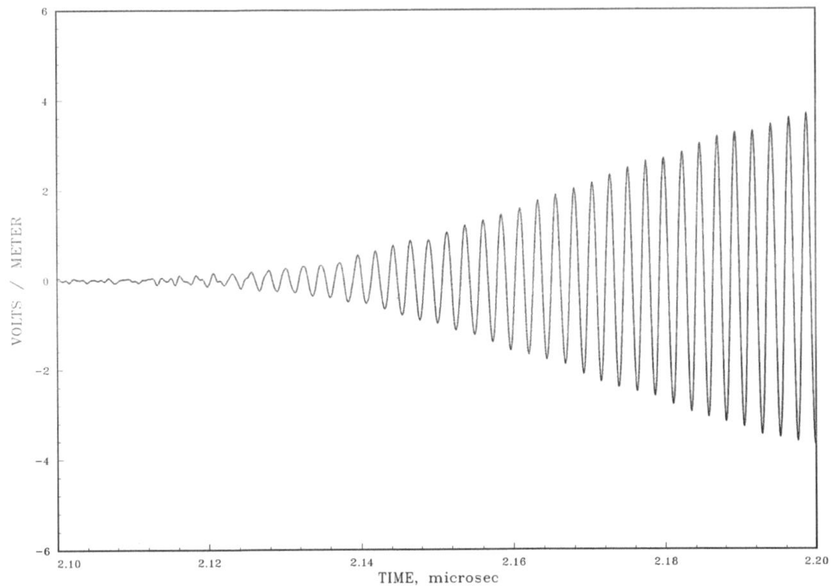

FIGURE 5-8 Single element buildup.

From these data, several conclusions may be drawn. First, none of the buildup or decay plots show anything that can be related to the time delay. Second, the buildup time is very much longer than the maximum delay time, for both array and reflector. Third, noise and interference were much worse for the D-dot measurements; all measurements discussed here were made with the TEM horn. Fourth, these results agree with the buildup calculations of Figures 5-7a and 5-7c, of the Air Force Phase IV waveform report; the discrete nature of the buildup can only be glimpsed at the very beginning, when the waveform has very small amplitude, and is heavily corrupted by noise and interference.

Next, delay in reflector antennas will be examined. Delay occurs in a reflector as a transmitted pulse appears first at the center of the reflector, then spreads out. The delay period continues until the farthest edge is excited, and the wave has reached a given angle off-axis (Hansen 1987; Hansen 1992) This delay for angle θ off-axis is:

$$\Delta t = \frac{D/\lambda}{2F}\left[\frac{\cos\theta}{8f/D} + \sin\theta\right].$$

Here D is reflector diameter, λ is wavelength, F is frequency, and f is focal length. For the SRI dish, $D = 150$ ft and $f/D = 0.42$. At center frequency of 435 MHz, and at an angle off-axis of 46 degrees, the maximum time delay is 70 nsec. This represents roughly 31 carrier cycles. Two transmitters were available for the SRI measurements. The wideband source had a rise time of 100 nsec; the narrow-band source had a rise time of 500 nsec. Figure 5-9, wideband data, is extracted from Figure 3-73c of the waveform report; only a small part of the buildup is shown.

Note the significant noise and interference before start of buildup, and its effect after buildup. Although the delay is 71 nsec, its effects are not discernable. Figure 5-10, from Figure 3-63e in the waveform report, shows buildup for the narrow-band dish.

Note that both interference and its effect on the rising waveform are reduced from Figure 5-9. It was stated by the measurement team that the interference at the SRI site was appreciable, and these graphs show this. In comparing Figure 5-6, PAVE PAWS buildup off-axis, with SRI buildup off-axis, the clear conclusion is that in neither can the delay effects be observed.

Turning now to the waveform decay, Figure 5-11 (from Figure 3-56b in the Air Force Phase IV wave form report) shows the array at wide angle.

Figure 5-12 (from Figure 3-53e in the Air Force Phase IV wave form report) shows the array near normal, where there are negligible time delays. Both show erratic decay; that at normal to the array face is probably less well behaved. In both cases the effects of interference are visible. Nothing in Figure 5-10 can be traced to an array discrete time delay.

For the reflector antenna, Figure 5-13 (from Figure 3-76c in the Air Force

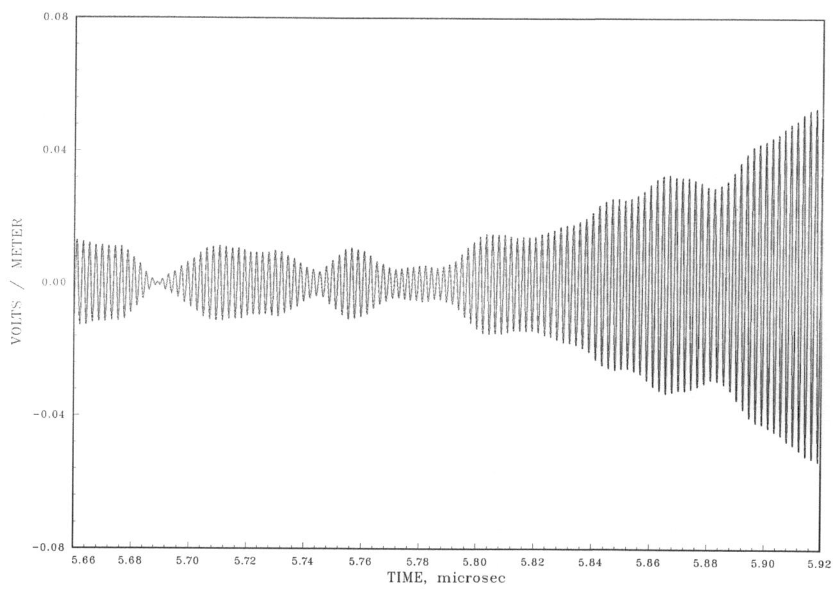

FIGURE 5-9 Dish buildup; 46 degrees off normal wide band.

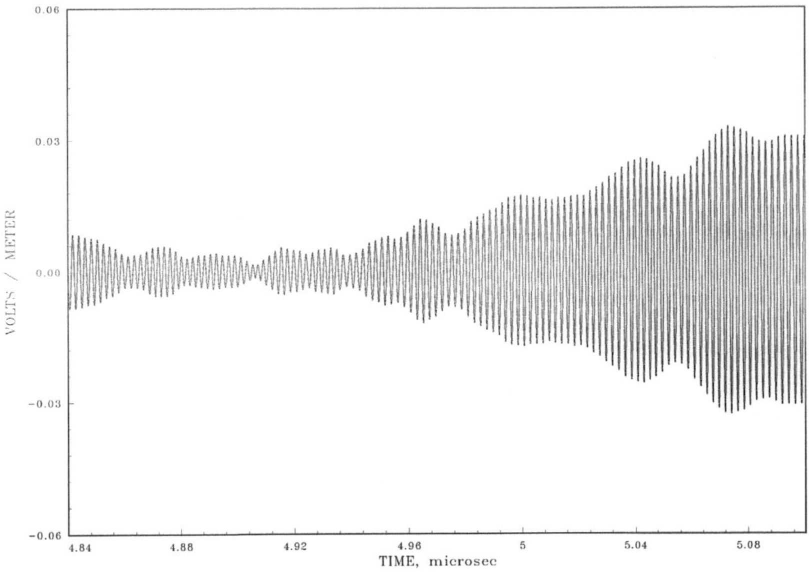

FIGURE 5-10 Dish buildup; 46 degrees off normal — narrow-band.

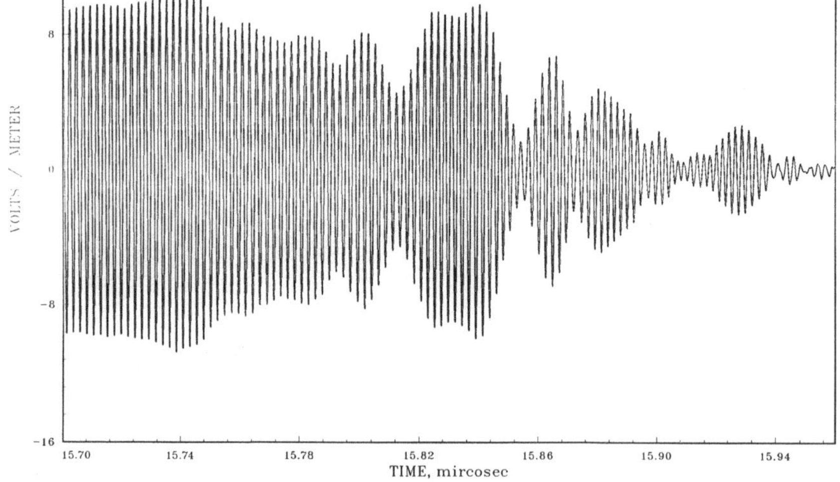

FIGURE 5-11 Array fall-off; 60 degrees off normal.

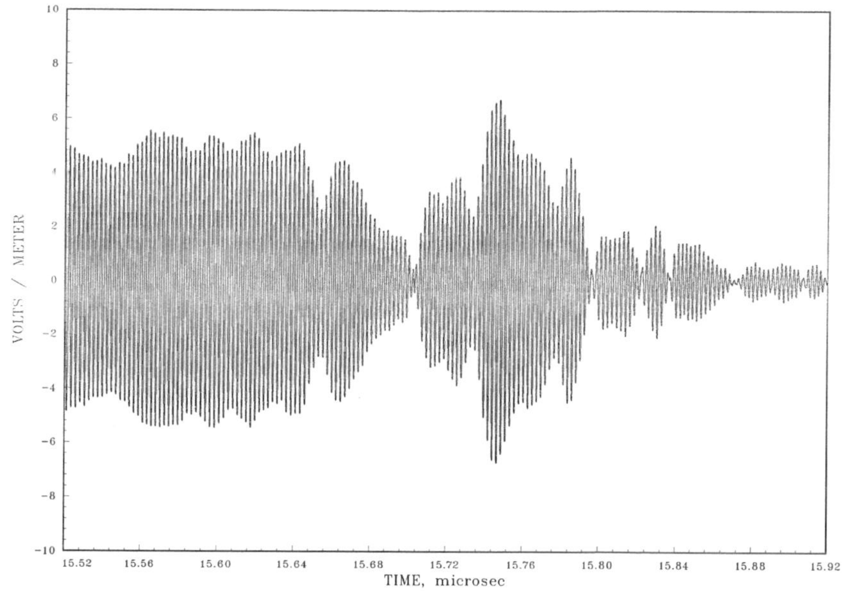

FIGURE 5-12 Array fall-off; near normal to array face.

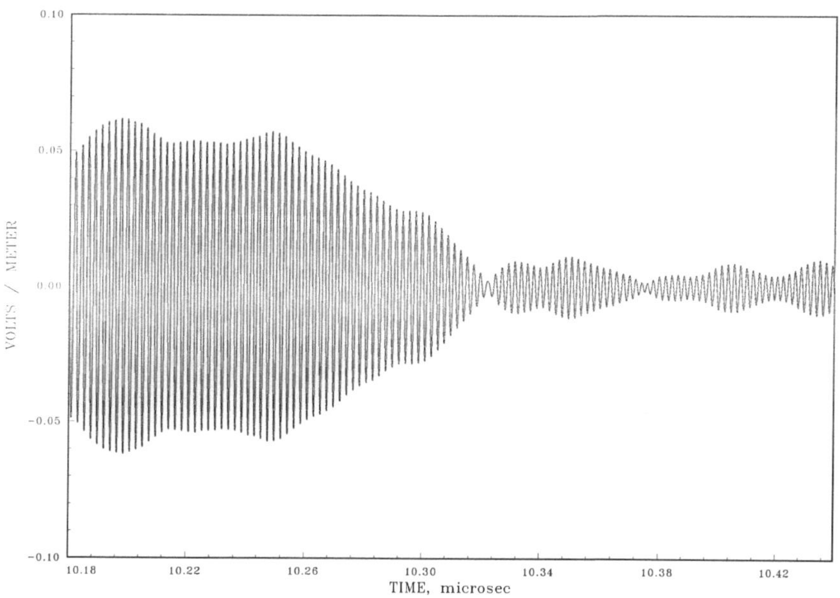

FIGURE 5-13 Dish fall-off; 46 degrees off normal.

Phase IV wave form report) shows the wideband decay. Again the effects of interference are apparent during the decay, and after. Note that for this case, the source rise time is 100 nsec. Again no effects of the reflector delay are visible. The narrow-band source decay graph is not shown; it decays much more slowly, due to the 0.5 μsec rise time, and the interference is correspondingly reduced.

PRECURSORS

Radar Electromagnetics

Another key argument by Dr. Albanese regarding PAVE PAWS was that precursors might be formed. If so, their deeper penetration into cellular material might be a factor to consider in the evaluation of possible health effects. In communications with the committee, Dr. Albanese recommended that the PAVE PAWS committee take advantage of the extensive technical literature on measurement and calculation of precursors. Accordingly, all available measurements are discussed; several Annexes are devoted to calculations.

The committee has identified the following questions regarding PAVE PAWS, biological tissues, and precursors. These questions are (1) does the radar radiation produce precursors in biological tissue or equivalent media; (2) does precursor formation depend upon signal bandwidth; (3) if precursors are produced, how do they decay with penetration distance? Definitive and reasonable answers to these questions have been developed, and are discussed below.

Central to all of these questions is bandwidth, which is critical here, as it is in all electronic and computer systems. Signal bandwidth is affected by both waveform rise time and pulse width. For PAVE PAWS the pulses are long (shortest pulse is 250 μsec); the bandwidth is primarily due to rise time. A continuous-wave (CW) signal may have a rapid rise time (roughly 1/4 of a cycle), but the CW signal has zero bandwidth. Conversely, phase-modulated signals such as QPSK may have wide bandwidth, but the signal is always on. It is wide bandwidth that is of concern for precursor formation.

Central to all of these questions are the electrical properties of the material the radar signal is traveling through. Precursors occur in anomalous dispersive media, that is, where there is a decrease in the permittivity (dielectric constant) with increasing frequency and where one or more absorption bands exist in the frequency range of the signal. See Annex 5-1 for data on dispersion in biological tissues.

A precursor is a modification of the incident waveform that arrives at a given depth in the dispersive medium before or during the incident waveform. Sommerfeld precursors arrive before, and tend to contain frequencies higher than the signal carrier frequency. Brillouin precursors arrive next, and tend to contain lower frequencies. In the medium, the incident waveform travels with the group velocity. The wave front of the Sommerfeld precursor travels with the speed of light in a vacuum; but this wave front is extremely small. The signal velocity, based on energy transport, is slower. However this signal velocity is larger than the group velocity, which is that of the incident waveform, accounting for the early apperance. In contrast, the Brillouin precursor has a signal velocity below the group velocity. Phase velocity has no meaning except for a monochromatic (single frequency) signal (Oughstun and Sherman, 1994).

Precursors may exist in a dispersive medium (dielectric constant/permittivity) and/or loss varying with frequency. Whether they are excited depends also upon the bandwidth of the incident signal. Wideband signals can produce precursors in highly dispersive media. But the variation in material parameters with frequency over a narrow-band signal does not allow observable precursors to form. This is substantiated by the Sandia National Laboratory calculations (see Annex 5-3).

The PAVE PAWS instantaneous bandwidth is never more than 5 MHz, although the carrier frequency changes as the beam moves between 420 MHz and 450 MHz. Thus this narrow bandwidth will not produce measurable precursors.

When precursors are created, the decay is a function of bandwidth of the applied waveform. Roberts (see Annex 5-3) has shown that a wideband signal will have significant spectral energy below the frequency of the carrier. Each of these spectral components decays exponentially in a linear lossy medium; if the wideband applied waveform has sufficient lower frequency energy, the resulting precursor will decay more slowly than the exponential decay of the carrier. Because the spectral components of lower frequencies decay more slowly with penetration distance, the mix (spectrum) changes with distance, resulting in a decay rate for the Brillouin precursor of approximately $1/\sqrt{R}$ (papers by Oughstun and by Roberts; see Annex 5-3). Note that all examples computed by Oughstun and colleagues utilize very wideband waveforms, typically bandwidths of 10 to 20 GHz or more, at optical (laser) frequencies. For PAVE PAWS with its very narrow bandwidth, the carrier frequency predominates, and all decay in a lossy media will be exponential at this frequency.

Waveform Slopes

Slope is the rate at which electric field is changing. Thus it may be measured in volts/meter per nanosecond. This is not to be confused with rise time, in nanoseconds, which is independent of field strength. At the September 16, 2003, PAVE PAWS committee meeting, Dr. Albanese presented charts on phase and slope estimation. The zero crossings of waveforms were examined to determine slopes. The slope between adjacent zero crossings changes at rates greater than 1 volt per meter per nanosecond have been postulated to be biologically significant by some individuals. However at the zero crossings, the PP signal is approaching zero; Jim Tomlin has shown in detail[3] that noise and interference, although small, can sig-

[3]Jim Tomlin, "Interior Delta Slopes and Noise," charts, notes, and attachment prepared for the PPPHSG.

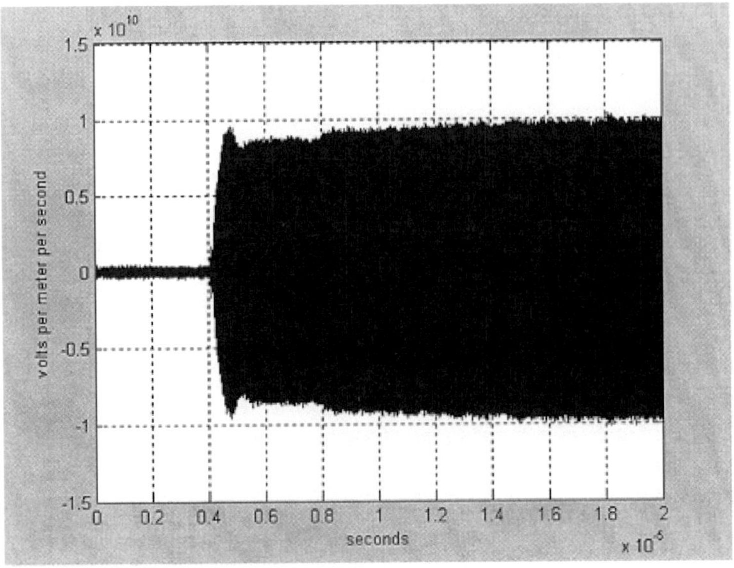

FIGURE 5-14 PAVE PAWS spatial variation at 500 meters. Used with permission from Richard Albanese.

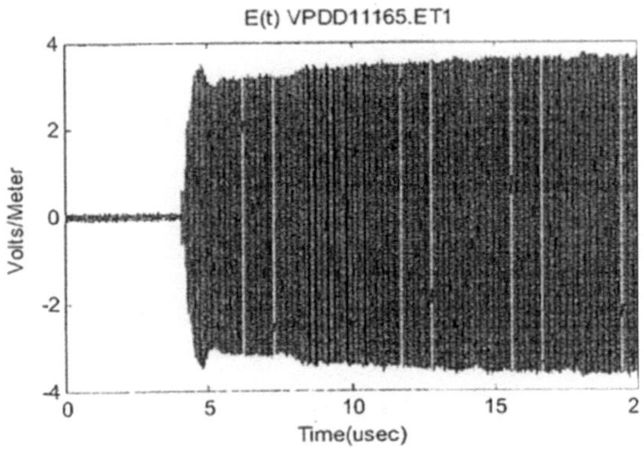

FIGURE 5-15 Leading edge 540 meters from PAVE PAWS.
A similar match occurs in FIGURE 5-16 (vpdd 11173.ed1) and FIGURE 5-17 (Figure 3-106h from the waveform report).

nificantly change the slopes because in the vicinity of a zero crossing the PP signal is smaller than the noise. As a result, the slopes and slope changes (delta slopes) calculated by Dr. Albanese are corrupted by interference. The spatial variation in Figure 5-14 matches almost exactly the leading edge of the related Phase IV waveform in Figure 5-15; Figure 5-14 should be volts/meter, as in Figure 5-15.

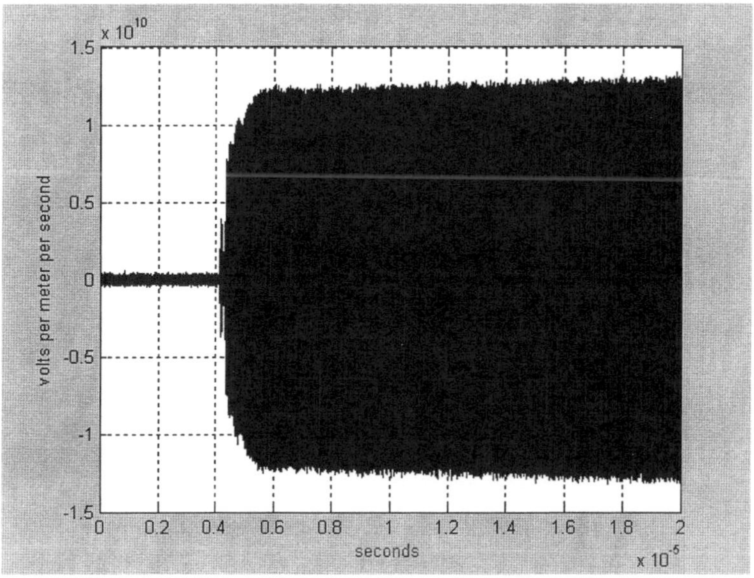

FIGURE 5-16 PAVE PAWS Spatial variation at 500 meters.

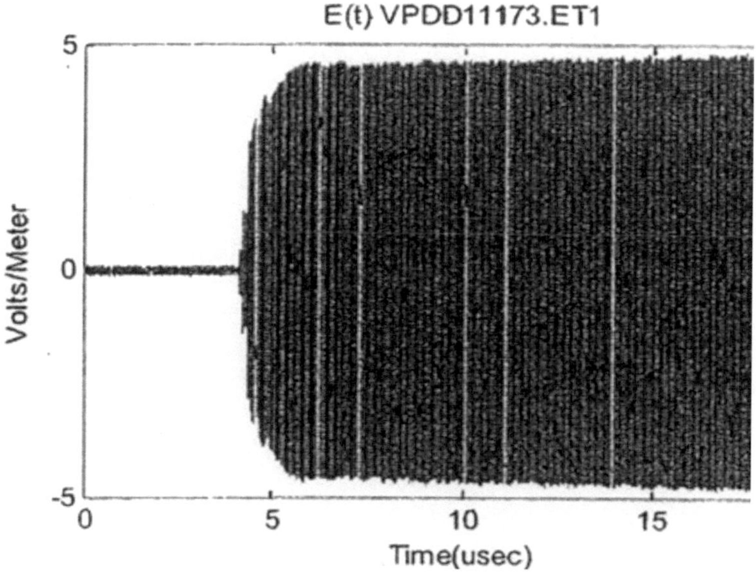

FIGURE 5-17 Leading edge, 540 meters from PAVE PAWS.

The "wiggles" in the Phase IV waveform envelopes are caused by interference, and these same wiggles are added to the PP carrier slope of 2.7 V/m/nsec (for a 1 V/m signal at 425 MHz). Slopes in Figure 5-18 (provided by Dr. Albanese as figure parl11090.ed1) are shown over a 400 nsec time, representing about 170 carrier cycles. The granularity maximum delay is 60 nsec; the waveform buildup is completely covered by Figure 5-18 but nothing attributable to the discrete delay is visible.

Jim Tomlin, for the PPHSG, has smoothed the waveform data, and has calculated and plotted delta slopes (see Figure 5-19). The upper curve, with right scale, shows the smoothed (two-point running average) of the noise and interference derivative for a record segment just prior to the arrival of the indicated PAVE PAWS pulse. The lower curve, with left scale, shows the delta slopes that would have been recorded due to this noise and interference had a pure 434.8 MHz non-phased sine wave signal been present. It is clear that the variations in delta slope are due to interference. All high-power radiating systems (TV, FM, radars) have

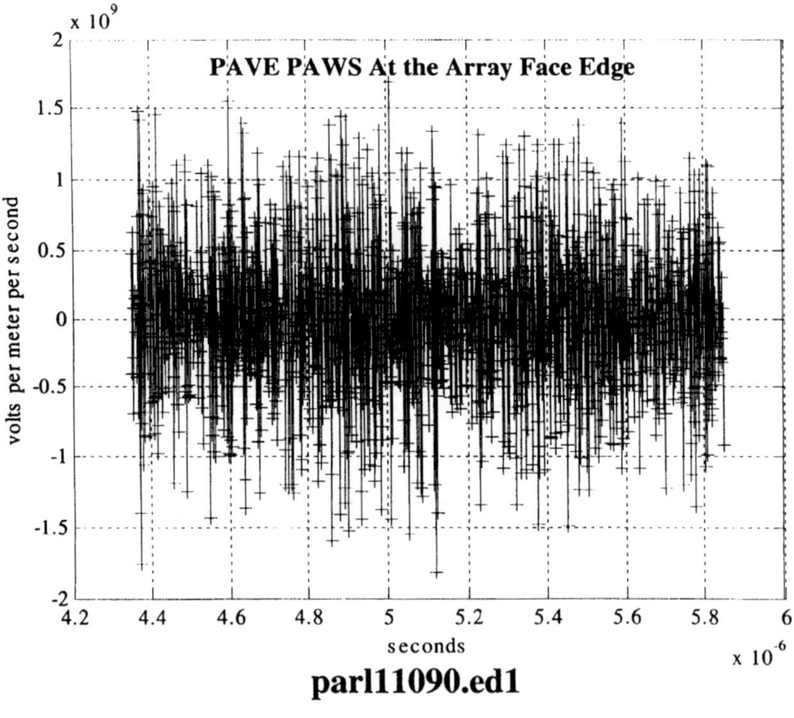

FIGURE 5-18 Slope differentials at adjacent zero points. Used by permission from Richard Albanese.

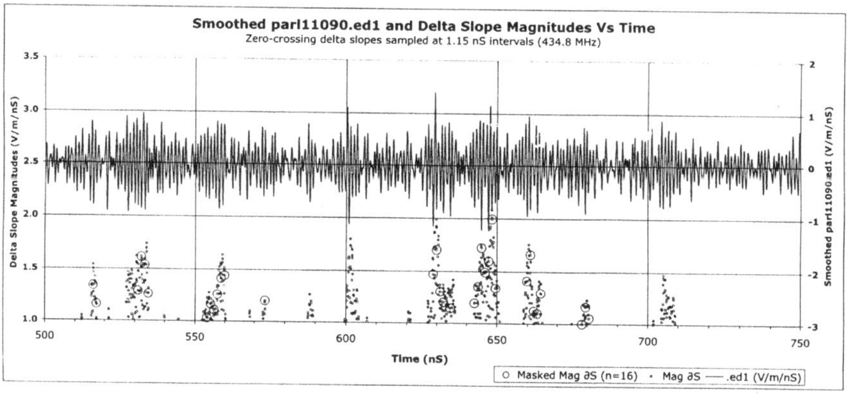

FIGURE 5-19 Smoothed waveform data and delta slope magnitudes vs time. Used with permission from Jim Tomlin.

slopes of several V/m/nsec, plus delta slopes induced by interferers. There is no reason to believe that PAVE PAWS effects are different. Tomlin has suggested additional measurements; the committee concludes the variations are due to interference and that further measurements are not needed.

Relation of Waveform Slopes to Precursers

As part of the committee's information-gathering effort, Dr. Albanese was asked by the committee what, in his opinion, the relationship was between slope (V/m/nsec) and precursor formation. He responded that the association was indicated in the paper by Albanese, Penn, and Medina (see Annex 5-3). A careful reading of this paper shows that Brillouin precursors occur when the medium is sufficiently dispersive, and when the bandwidth is wide. Their calculations used a convenient applied field of 1 V/m, and the precursor levels were related to this. The system is linear; it is clear that a hypothetical applied field of 1 mV/m would produce a precursor with amplitude 1/1000 of the precursor in the paper. This paper shows indirectly the association between signal rise time (in nsec) and precursor formation. Slope (V/m/nsec) is not associated therein. It is important to distinguish between rise time, in nsec, and slope, in V/m/nsec.

Discussion of Extant Measurements

The committee has located only three measurements that showed precursors. First, circa 1967, as thesis work at NYU-Bronx; second, circa 1986, at Physics International; third, calculations and measurements made at the Naval Surface Warfare Center in the 1990s, but published in 2001. In addition, circa 1990, calculations and measurements were made at the University of Michigan but not published, and later briefly extracted. Further measurements were made by AFRL-DED contractors on water, sand, and concrete. The Michigan and AFRL measurements did not show precursors.

Pleshko (see Annex 5-2)

Graduate student Peter Pleshko, under the supervision of Professor Istvan Palocz, designed experiments to produce precursors in the microwave region, and took measurements accordingly, all as a Ph.D. project at New York University-Bronx. Their paper describes and discusses the three experiments they designed to demonstrate precursors.

Precursors were observed, but in all three experiments the bandwidths were wide: the smallest bandwidth was roughly 500 MHz. Thus, there is minimal relevance to PAVE PAWS.

Physics International Report "Dispersive Pulse Propagation" (see Annex 5-2)

Physics International (PI) experiments replicated one of the Pleshko experiments, using a coaxial test fixture for additional measurements and also open-tank measurements. Measuring and test equipment for short-pulse wideband were somewhat primitive (circa 1987) compared with current capabilities. The source signals were of poor quality: the source pulse had high pre- and post-pulses. The coaxial test cell was designed for a 50 ohm match with air; the water, salt, and sugar solutions produced major mismatches resulting in quite apparent ringing. All of the measurements were wideband, and had frequency spectra centered above 1000 MHz, where dielectric constant ε and conductivity σ changes with frequency are significant. Precursors were observed only in those few wideband experiments that were designed to create strong dispersion (waveguide and Goubau line). No discernable precursors were observed in test cell and liquid tank experiments, even though bandwidths were wide and frequency spectra were in the region of changing dielectric properties.

Stoudt, Peterkin, and Hankla, NSWC report "Transient RF and Microwave Pulse Propagation in a Debye Medium (Water)" (see Annex 5-2)

The measurements utilized a coaxial line test cell filled with re-circulating deionized water. Waveforms were wideband; low-frequency precursors were measured. Agreement with calculated data was excellent. These measurements used well-designed equipment with realistic media, and were of high quality.

Mourou and Williamson (see Barrett in Annex 5-2)

The paper by Barrett reproduces a small set of the measurements made at the University of Michigan by Mourou and Williamson. The Mourou and Williamson work has not been published. Measurements were made using microwave stripline with beta-alumina dielectric. A wideband waveform at 8 GHz was used, but the apparatus introduced significant reflections. Nonetheless, no signals that could be identified as precursors were observed.

Farr and Frost, Measurement Notes 49 and 52, AFRL-DED (see Annex 5-2)

The purpose of the measurements was determination of dielectric properties of water, sand, and concrete. A coaxial transmission line was used, with an impulse waveform with spectral content from DC to over 20 GHz. Significant pulse broadening and attenuation were observed, but no overt precursors appeared.

Discussion of Calculations

Extensive calculations have been published by Oughstun and colleagues, in a book and numerous papers (see Annexes 5-3 and 5-5). All of these use wideband waveforms, typically at laser (infrared) frequencies. Classic precursor waveforms are produced, but the vital contribution of bandwidth is not discussed.

Sandia Report "Calculations of Precursor Propagation in Dispersive Dielectries" (see Annex 5-3)

Bacon calculated propagation of waveforms into pure water at 435 MHz, for two bandwidths: 200 MHz, and 10 MHz. The 200 MHz bandwidth waveforms showed precursors, while the 10 MHz bandwidth waveforms did not. The narrow-band signal also showed a slow buildup (and presumably decay). The wideband waveforms decay at the carrier rate for several meters, then more slowly. However the band-limited signal decays at the carrier rate at all penetration distances. PAVE PAWS has band-limiting filters in the exciter units, so these 10 MHz results are expected to represent the radar (see annex 3).

These calculations were buttressed by Roberts, who showed that decay slower than carrier exponential requires significant spectral energy below the carrier frequency, i.e., wideband. As mentioned in Radar Electromagnetics, the spread of spectral energy allows the precursors to decay as $1/\sqrt{R}$ at distances where the main signal is negligible.

Moten and Others (see Annex 5-3)

Moten and colleagues calculated that a wideband waveform produces precursors, but stated that narrow band systems would not generate precursors (see Annex 5-3).

In several papers, Albanese and colleagues calculated precursors, but all for wideband signals. No attempt was made to correlate precursor generation with either bandwidth, or dispersive characteristics of the medium with frequency.

Mokole and Samaddar (see Annex 5-3)

Mokole and Samaddar calculated the behavior of a laser pulse (very wideband) propagating into a semi-infinite medium. Strong precursors were predicted.

Dvorak (see Annex 5-3)

Dvorak, using exotic functions, calculated propagation in a waveguide, thus verifying the prior experimental work of Pleshko.

Discussion of Annexes

The annexes contain a citation and review of each journal paper that appears relevant. They are grouped into precursor measurements, precursor calculations, related papers on precursors, pulse propagation in lossy media, and publications by Professors Oughstun and others.

SUMMARY

The PAVE PAWS phased array contains a large number of elements per face (1792), all of which are turned on simultaneously. Bandwidth is very narrow, 5 MHz at most; the signal is heavily filtered to reduce interference. Waveform buildup is slow (about 200 nsec) due to the signal bandwidth and to the amplifier characteristics. Waveform slopes, in V/m/nsec, are as expected, except in the vicinity of zero crossings; there, noise and interfering signals control slopes, and the PAVE PAWS signals are negligible. Radial fields as such are not present; the longitudinal field components observed were produced by the dipole 20 degree tilt, and multipath fields.

At angles away from the array axis, time delay exists between the signal from the closest dipole and the farthest dipole. It is a maximum of about 60 nsec. Reflector (dish) antennas have a similar delay at wide angles; the same-size array and reflector have roughly the same delay. For PAVE PAWS the peak delay is roughly 1/3 of the buildup time. Due to the large number of active elements per face, and due to the pseudo-random distances from each element to an observer, the signal buildup is smooth as shown by detailed calculations. In none of the Phase IV measured data is it possible to discern the delay effects. The committee's conclusion is that narrow-band phased-array radiation is just like narrow-band reflector radiation.

At wide angles from the array axis and from the radar beam, the sidelobe level is extremely low. This low sidelobe level results from the destructive summation of 1792 contributions. Turnoff and decay of the Class C amplifiers is somewhat irregular, which sometimes results in decay transients above the normal low sidelobe level. Amplitude of these spikes is still much below the allowable power density; this, with their short duration, indicates a negligible effect.

Precursors are transient signals that may appear before, during, and after the signal, in a dispersive medium such as water or cells. None propagate faster than light. Those that have been considered as possibly important have significant low-frequency spectral energy, that is, important frequency components below the carrier frequency. There are extensive calculations, and several measurements, that show that wideband signals produce observable precursors. Typical precursor wide bandwidth is 10,000 MHz, a factor of 2000 times the PAVE PAWS bandwidth. Each frequency component of the wideband waveform decays exponentially in the medium. After the signal has experienced a large attenuation, the precursor waveform may decay algebraically. Precursor formation is related to signal bandwidth (buildup time) but not to signal level (V/m). The criterion of 1 V/m/nsec has no meaning for precursor formation. As discussed under Precursors: Radar Electromagnetics, rise time and bandwidth are independent of field strength in V/m.

For the PAVE PAWS bandwidth no measurable precursors were found in the several experiments. All calculations and measurements conclude that narrowband phased-array radiation effects are not distinguishable from those of reflector antennas. This may also be true for wideband phased arrays, but no relevant data are available.

REFERENCES

AFRL (Air Force Research Laboratory). 2003. Phase IV—Time-Domain Waveform Characterization Measurements of the PAVE PAWS Radar. Kirtland Air Force Base, Albuquerque, NM.

Baños, A. 1966. Dipole Radiation in the Presence of a Conducting Halfspace. New York: Pergamon Press.

Brillouin, L. 1962. Wave Propagation and Group Velocity. St. Louis, MO: Academic Press.

Collin, R.E. 2004. Hertzian dipole radiating over a lossy or sea: Some early and late 20th century controversies. IEEE Antennas and Propagation Magazine 46:64-79.

Hansen, R.C. 1987. Short pulse excitation of reflector antennas. P IEEE 134: 557-559.

Hansen, R.C. and E. Kramer. 1992. Correction to short-pulse excitation of reflector antennas. P IEEE 139:119-120.

Norton, K.A. 1937. Space and surface waves in radio propagation. Phys Rev 132-133.

Jordan, E.C. and Balmain, K.G. 1968. Electromagnetic Waves and Radiating Systems. Prentice Hall.

Oughstun, K.E. and G.C. Sherman. 1994. Electromagnetic Pulse Propagation in Causal Dielectrics, Springer Series on Wave Phenomena, vol. 16. Berlin-Heidelberg: Springer-Verlag.

Somerfeld, A. 1914. Above the propagation of light in dispersive media. Ann Physik 44:177-202.

ANNEX 5-1: DISPERSION IN BIOLOGICAL TISSUES

In biological tissues, three distinct dispersion regimes have been defined. The lowest frequency dispersion, referred to as the α dispersion, is very large, and occurs near 10 Hz to 100 Hz, and is believed to be associated with large biological structures such as cell bodies. The second dispersion (referred to as the β dispersion) occurs over the range of approximately 100 kHz to 100 MHz. It reflects the dipolar characteristics of the molecular constituents of the tissue, and terminates at the permittivity of water (approximately 80 farads/meter). The third dispersion (γ), is associated with the relaxation time of water and occurs in the low GHz range. Correspondingly, in the frequency range of the PAVE PAWS signal, the dielectric permittivity of biological tissue is very near to that of water, and the dispersion is very low. While the dispersion (δε/δf) depends on both the specific tissue and frequency range, dispersions in biological tissues near 500 MHz are in the range of only 1 to 10 farad/meter per 100 MHz. This is a very small dispersion. Nonetheless, whenever the frequency spectrum of an applied waveform encompasses a region of anomalous dispersion, precursors may be formed. If the bandwidth is narrow, and the material parameters change little if at all over the frequency range of interest, precursors will have negligible amplitude. An examination of the dielectric properties of the white matter of brain tissue, for example, shows that the permittivity changes at a relatively constant rate of 0.5 decade of permittivity/decade of frequency, from a value of 10,000 at 10 KHz, to a value of 30 at 1 GHz. The PAVE PAWS instantaneous bandwidth is never more than 5 MHz, although the carrier frequency changes as the beam moves between 420 MHz and 450 MHz. Thus this narrow bandwidth signal is not expected to produce measurable precursors.

ANNEX 5-2: PAPERS ON PRECURSOR MEASUREMENTS

1. P. Pleshko and I. Palocz, "Experimental Observations of Sommerfeld and Brillouin Precursors in the Microwave Domain," ***Phys. Rev. Letters*****, Vol. 22, 2 June 1967, 1201-1204**

Graduate student Peter Pleshko, under the supervision of Professor Istvan Palocz, designed experiments to produce precursors in the microwave region, and took measurements accordingly, all as a Ph.D. project at New York University-Bronx. This paper describes and discusses the experiments they designed to demonstrate precursors.

The first experiment utilized a coaxial transmission line filled with ferrite material, which was magnetized by an applied external magnetic field. This simulated the first branch of the dispersion curve. A pulse of sine waves at a carrier frequency of 625 MHz was applied at one end of a long coax, and the waveform was observed at the other end. With an applied magnetic field of 100 gauss (earth field 0.3 to 0.6 gauss), the Brillouin precursor was visible but the main signal was not. With a 200 gauss field the Brillouin precursor, which looks like one cycle with a slightly lower frequency than the carrier, was followed by the pulse train. The precursor amplitude was nearly twice that of the pulse train.

The second experiment was aimed at simulating the second branch of the dispersion curve. A waveguide with a cutoff frequency of 4290 MHz was used; the guide length was 6 ft. A pulser source producing a 100 picosecond single pulse with fast rise time was used. A Sommerfeld precursor with a frequency of roughly 15,000 MHz was observed, but the amplitude was small compared to the main signal; these precursors are of higher frequency.

The third experiment was also second branch, and used a Goubau surface-wave transmission line consisting of RG-8 coaxial cable with the outer braid removed. It was then a wire with a dielectric sheath. Horn feeds were used. Again the 100 picosecond pulser with 30 picosecond rise time was the source. The Brillouin (low frequency) precursor was observed; the pulse was not shown.

In all three experiments described above the bandwidths were large: in experiment one, 350 Mhz or 315%; experiments two and three, 350 MHz. These large bandwidth[s] are critical, as signals occupying large bandwidth can experience significant dispersion (dispersion is a change of parameters with frequency), and have major energy at low frequencies. In spite of the deliberately introduced large dispersion, the high frequency (Sommerfeld) precursor was very small. The amplitude of the low frequency (Brillouin) precursor was not compared to the pulse amplitude. The large precursors calculated for the optical region by Oughstun did not develop, but the optical bandwidth and dispersion are much larger, and the wavelength is very much shorter than at microwave frequencies.

2. R.R. Smith, Dispersive Pulse Propagation, report PITR 3341, Physics International Co., prepared for USAFSAM/RZM, Dec. 1988

The principal investigator performed three experiments to determine if precursors could be generated by microwave frequencies.

The first PI experiment replicated the Goubau surface-wave transmission-line experiment of Pleshko and Palocz. A step-input function was used, with a rise time of roughly 300 picoseconds. The Brillouin precursor was observed as an overshoot of roughly 100% with rapid decay. Signal bandwidth was approximately 6700 MHz.

The next experiment simulated muscle tissue in a coaxial test fixture filled with distilled water. A single-step function with a rise time of roughly 500 picoseconds was the applied signal. For the frequency spectrum of the waveform, roughly 4000 MHz wide, distilled water, if sufficiently pure, has dielectric constant and conductivity similar to those of muscle tissue. The results showed severe waveform distortion and ringing (steps) as illustrated in trace 906D of Figure 8 of the report. Unfortunately the coaxial line was designed for 50 ohms impedance (industry standard for test equipment) when air filled. When water or ionic solution fills the test cell, the impedance is badly mismatched, resulting in the serious ringing observed.

Another experiment used a pulse generator that supplied a bipolar pulse (a positive pulse followed by a negative pulse), again into distilled water. This pulse was of poor quality, having itself a small precursor and a small postcursor, along with a negative pulse much lower in amplitude and wider in time, than the positive pulse. Results from the test cell showed considerable ringing, no doubt due to the impedance mismatch, but no evidence of Sommerfeld or Brillouin precursors. Measurements next used water with 1.4% salt, with the step-function signal source. Results showed severe waveform distortion but no evidence of precursors. With the bipolar-pulse signal, again results were distorted but no precursors were present. Use of a sugar/salt mixture with water produced essentially the same results: distortion but no precursors. Coaxial test fixture measurements were contaminated by ringing due to impedance mismatch, and by poor quality source signals. No Sommerfeld or Brillouin precursors were observed.

Open-tank measurements used a waveguide as an illuminator, with Set One with cut-off at 910 MHz and Set Two with cutoff at 1320 MHz. The signal was tripolar (–,+,–) but was of poor quality with ringing. Results did not show any precursors for several salt and sugar solutions. Finally, as the signal propagates into the lossy solution, the higher frequency components are attenuated more, as expected. The source signal was of very poor quality, and there were internal and external reflections, resulting in extraneous signals (ringing). No precursors were observed.

Short-pulse, wideband measuring and test equipment was somewhat primitive (circa 1987) compared with current capabilities. The source signals were of poor quality. The coaxial test cell was designed for a 50 ohm match with air; the water, salt, and sugar solutions produced major mismatches with resulting quite apparent ringing. These reflections tended to mask the changes in waveform produced by the medium. All of the measurements were wideband and had frequency spectra centered above 1000 MHz, where dielectric constant and conductivity ? changes with frequency are significant. Precursors were observed only in those few experiments that used wide bandwidth and deliberately created strong dispersion (waveguide and Goubau line). No discernible precursors were observed in test-cell and liquid-tank experiments, even though bandwidths were wide and frequency spectra were in the area of changing dielectric properties.

3. T.W. Barrett, "Energy Transfer and Propagation and the Dielectrics of Materials: Transient Versus Steady State Effects," *Proc. Ultra-wideband Radar Symp.*, Los Alamos, CRC Press, 1990, 1-20

In this paper, the author gives a brief review of the saddle-point approximations used in calculating transient response of dispersive materials. A much more thorough account is given by Oughstun and Sherman. The paper reproduces some of the unpublished measurements made circa 1990 at the Ultrafast Science Laboratory of the University of Michigan by Professor Mourou and Dr. Williamson. A more extensive set of the raw measured data has been made available by Dr. Whittaker of the University of Michigan. The relevant measurements utilized strip line with either air or beta-alumina as the dielectric. The frequency was 8 GHz, and waveforms of one to eight cycles were used. Unfortunately, the waveform generator and the detector/receiver together produced several significant reflections (echoes) in the reference air line results. Reflected waveforms appeared at roughly four and eight cycles. Using a length of dielectric of 1 inch, both single-cycle and eight-cycle incident waveforms showed a distorted waveform due to the echoes, but no precursors. With a dielectric length of 4 inches, the carrier waveform was absent due to attenuation, but the transmitted waveform did not look like any calculated precursors. In all of these experiments, the bandwidths were large, and extended to frequencies well below the carrier frequency.

4. D.C. Stoudt, F.E. Peterkin, and B.J. Hankla, "Transient RF and Microwave Pulse Propagation in a Debye Medium (Water)," NSWC report JPOSTC-CRF-005-03, Dahlgreen, VA, Sept. 2001

The test cell was a 1.4 m-long coaxial transmission line with recirculated deionized water. A tapered dielectric plug matched the 33 ohm line (when filled) to the 50 ohm test equipment. A short was placed at the far end of the line, so that the signal would be reflected. At the feed end, a tee junction separated out the return signal. Round trip delay was 82 nanoseconds. Equipment included a Gigatronics 7100 signal generator and an HP 54720A Digitizing Oscilloscope.

A 10-cycle pulse at 1 GHz was the test signal. This signal showed two small amplitude residual cycles at the end of the signal, but the input signal quality was good. The output signal, after 2.76 m of water travel, showed only two precursors, no main signal. The fore and aft precursors were of opposite polarity, and closely matched the calculated values. A 10.5-cycle 1 GHz waveform showed similar results, and again agreed with calculations. In this case, because of the half cycle, both precursors are of the same polarity.

Two 6-cycle 400 MHz waveforms were used as exciters. The square-wave modulated pulse, after 2.76 m of water, showed precursor amplitude roughly twice the main signal amplitude. Again, calculations and measurements agree very closely. It is worth noting that the calculations for this waveform after 1.5 m of water showed only very slight precursor effects. Trapezoidal modulations of the 10 cycles of carrier were also tested. Only small precursor effects appeared after 2.76 m of water.

These measurements were carefully designed and conducted, with attention paid to impedance matching, use of time delays to separate input and output signals, and data recording. Thus these are the definitive measurements utilizing representative media.

5. E.G. Farr and C.A. Frost, "Time Domain Measurement of the Dielectric Properties of Water in a Coaxial Test Fixture," Measurement Notes 49, Dec. 1996; "Impulse Propagation Measurements of the Dielectric Properties of Water, Dry Sand, Moist Sand and Concrete, "Measurement Notes 52, AFRL Directed Energy Directorate, Albuquerque, NM.

An oversized coaxial transmission line, with the sample occupying part of the line length, was used. The signal was a wideband impulse (short pulse), with major spectral energy from DC to over 10 GHz. The line was used in both the transmission and the reflection modes. Unfortunately, the line was matched with air for 50 ohms, which produced significant mismatch at the sample interfaces. Thus, the results were contaminated with significant ringing. In passing through the distilled water sample, the impulse waveform was significantly attenuated and broadened. The resulting waveforms did not show any of the typical oscillations commonly seen in precursor waveforms. Although the time domain instrumentation was up to date and of high quality, the results were degraded by the design of the coaxial test chamber.

ANNEX 5-3: PAPERS ON PRECURSOR CALCULATIONS

1. L. Brillouin, *Wave Propagation and Group Velocity*, Academic Press, 1960

This most interesting, but hard to find, book is a retrospective of the work by Sommerfeld and Brillouin on wave velocities and precursors. A translation (from German) of the classic 1914 paper by Sommerfeld is included, along with a translation (from French) of the classic 1914 paper by Brillouin; both concern propagation in dispersive media and saddle-point asymptotics. In the other chapters, Brillouin discusses several waveform velocities: group velocity, signal velocity, and energy-transfer velocity. His calculations of precursors are less useful, due to errors, as observed by Oughstun and others. Both Sommerfeld and Brillouin used the term "forerunners" rather than precursors; use of "precursors" apparently found favor relatively recently.

2. K.E. Oughstun and G.C. Sherman, *Electromagnetic Pulse Propagation in Causal Dielectric*, Springer, 1997

This treatise shows in detail the difficulties in obtaining approximate solutions to wave propagation in dispersive media. Crudely, the signal is represented in the complex phase plane by a pole, and the precursors by branch points. Elaborate asymptotics are used to obtain the propagated waveforms. Unfortunately (for PAVE PAWS relevancy), all calculations are at infrared wavelengths, where the frequency is roughly 10,000,000 times higher than that of PAVE PAWS. All waveforms are either single cycle or a few cycles. For these very wideband signals, precursor formation is calculated. Further, the work is mathematically oriented, and has little physics interpretation. In particular, there is no discussion of the crucial effect of bandwidth.

Oughstun and colleagues have published many informative papers; almost all concern laser (IR) waveforms. Even their papers in *Radio Science* are on optical (IR) waveforms.

3. T.M. Roberts, "Radiated Pulses Decay Exponentially in Materials in the Far Fields of Antennas," *Electronics Ltrs.*, Vol. 38, 4 July 2002, 679-680

This salient paper shows that low-frequency spectral components of a waveform can produce a decay rate in dispersive material that is lower than that of the carrier frequency. In a linear medium, and at low power levels, tissue is linear and all spectral components decay exponentially. See also:

4. T.M. Roberts and P.G. Petropoulos, "Asymptotics and Energy Estimates for Electromagnetic Pulses in Dispersive Media," *J. Opt. Soc. Amer. A*, Vol. 13, June 1996, 1204-1217 and "Addendum," *J. Opt. Soc. Amer. A*, Vol. 16, 1999, 2799-2800

5. T.M. Roberts, "Measured and Predicted Behavior of Pulses in Debye- and Lorenz-Type Materials." *Trans. IEEE* AP-52 Jan. 2004. 310-314.

Asymptotic calculations are shown to validate precursor measurements made using liquid-filled coax cables by AFRL contractors (see Annex 2). Results show algebraic decay after significant attenuation has occurred. Signal bandwidth was approximately 10 GHz; the widest PAVE PAWS bandwidth is 5 MHz, or 2000 times smaller. PAVE PAWS precursors will be too small to be measurable.

6. L.D. Bacon, "Calculations of Precursor Propagation in Dispersive Dielectries," SAND 2003-3040, Aug. 2003, Sandia National Laboratories

Dr. Larry Bacon of Sandia National Laboratories has calculated waveform propagation in lossy dispersive media. These cases utilized a carrier frequency of 435 MHz, which is in the middle of the PAVE PAWS band. Pure water was the medium.

He used a narrow-band (1 microsecond) pulse train, and a wideband (10 nanosecond) pulse train. Note that the narrow-band pulse has about the same bandwidth as the widest PAVE PAWS waveform. Each pulse train was utilized without, and with, a filter. The 1 microsecond waveform used a 10 MHz wide filter, while the 10 nanosecond waveform used a 200 MHz wide filter. The two filter bandwidths are shown in Figure 5-A3.1.

The wideband unfiltered waveform displayed at 1meter depth in the medium a precursor, small compared to the main signal; at 5-meter depth, the precursor was larger than the signal but very small compared to the signal at 1 meter. When the wideband filter was applied, no precursors were visible.

The narrow-band unfiltered waveform displayed at 1-meter depth in the medium a very small precursor; at 5-meter depth the precursor was again larger than the signal but very small compared to that at 1 meter. With the filter applied,

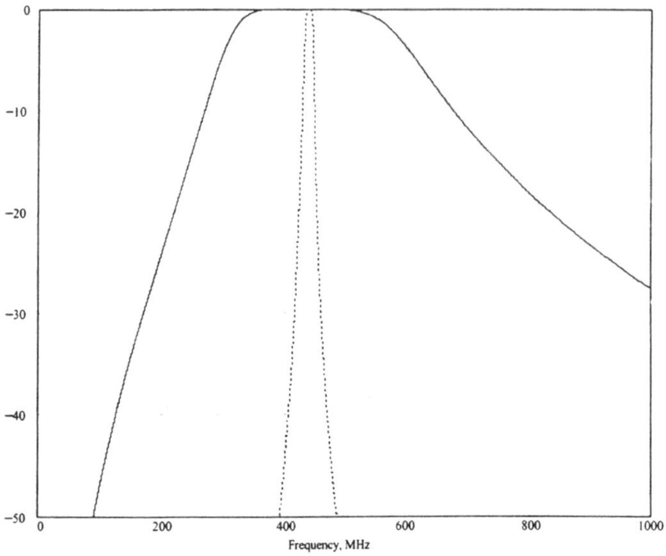

FIGURE 5-A3.1 10 MHz (dotted) and 200 MHz (solid) bandpass filter frequency responses.

which represents the PAVE PAWS radar, a slow buildup was observed at all depths, as seen in Figure 5-A3.2. No precursors were evident. However, at great depth in the medium, a precursor is visible but its amplitude is below the incident amplitude by a factor of roughly 100 million. Such small values are way below the natural noise level, hence are not measurable. Because of the narrow bandwidth of the 1 microsecond waveform, all spectral (Fourier) components decay essentially exponentially, producing the classic "skin depth."

Figure 5-A3.3 shows waveform decay over a 20-meter depth; this waveform decays exponentially just like all narrow-band waveforms. Exponential decay is linear on a logarithmic scale (decibels).

In summary, wideband waveforms do produce observable precursors. Narrowband waveforms, such as those of PAVE PAWS, do not produce measurable precursors.

7. D.C. Stoudt, F.E. Peterkin, and B.J. Hankla, "Transient RF and Microwave Pulse Propagation in a Debye Medium (Water)," NSWC report JPSTC CRF-005-03, Dahlgren, VA, Sept. 2001

This work concerns frequencies below 2 GHz, where water shows a nearly constant permittivity but a varying conductivity with frequency. Accordingly, group and signal velocity dispersion can be neglected, and the Debye dispersion model is adequate. The propagation of waveform is analyzed using Fourier Trans-

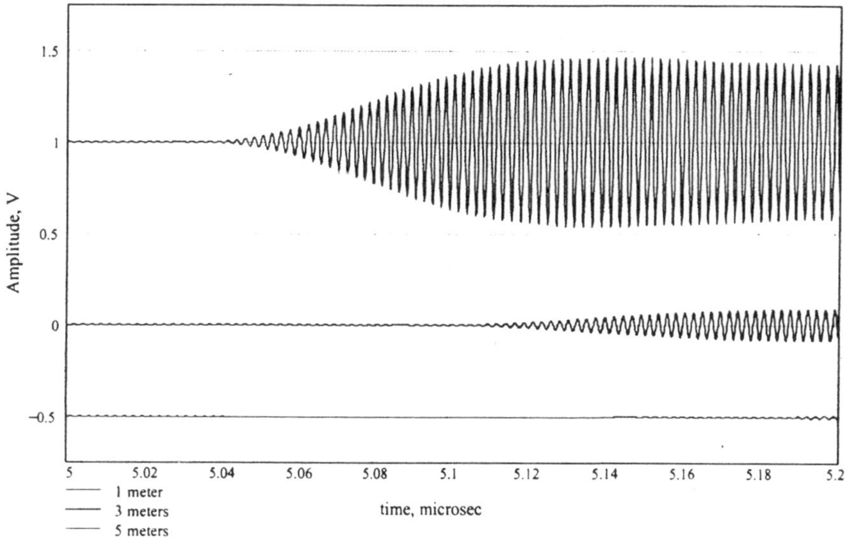

FIGURE 5-A3.2 Leading edge of 1 μsec signal at various depths; baselines are offset for clarity.

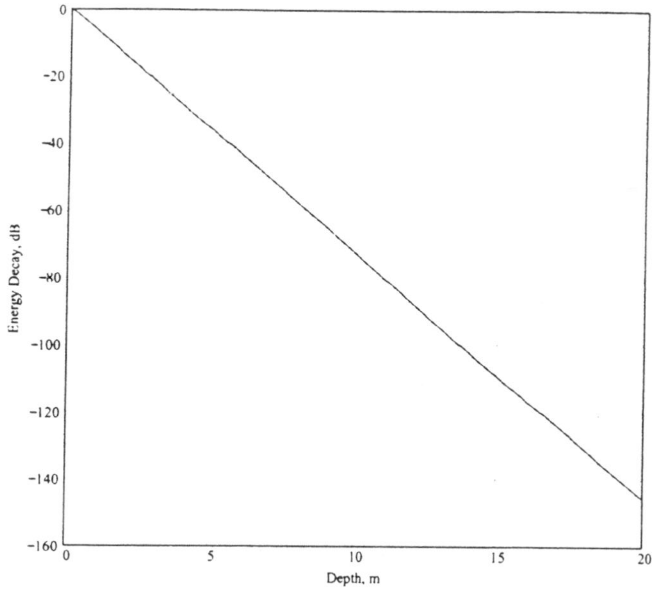

FIGURE 5-A3.3 Energy decay of 1 μsec signal, with 10 MHz filtering.

forms; the resulting frequency spectra are propagated through the lossy media, then inverse transformed into a waveform. Waveforms used include an ideal 10-cycle, 1 GHz pulse, a 6-cycle, 400 MHz pulse, and these two with a more practical trapezoidal envelope on the 1 GHz waveform. These calculations agree with those of Albanese (see reference 13, this section).

For the square-wave modulated 1 GHz signal, at a depth of 75 cm (2.5 ft.), the fore and aft precursors are much larger than the main 10-cycle waveform. At the much shorter depths representative of humans, the waveform would be much larger than the precursor. When the modulation is trapezoidal, the precursors are only slightly larger than the main signal, at the same 75 cm depth. Bandwidth of both signals is roughly 200 MHz. For depths so large that the main signal has disappeared due to strong attenuation, the precursor decays approximately as $1/\sqrt{R}$, thus validating the one-term series approximation of Oughstun.

The 6-cycle, 400 MHz signal, with ideal square-wave modulation (instant turn on), shows only a precursor slightly larger than the main signal, at a depth of 1.5 m (4.9 ft). Bandwidth is roughly 130 MHz.

These calculations show that for 400 MHz, even with the large bandwidth associated with the 6-cycle waveform, precursors are negligible at the limited tissue depth available in humans.

8. K. Moten, C.H. Durney, and T.G. Stockham, "Electromagnetic Pulse Propagation in Dispersive Planar Dielectrics," ***Bioelectromagnetics,*** **Vol. 10, 1989, 35-49**

This paper investigates a plane-wave pulse-train incident on a lossy dispersive slab, using Fast Fourier Transform techniques. Calculations show that wideband pulses can produce significant precursors. However, "the narrower the bandwidth of the pulse train, the less the difference between the pulse response and CW response." And, "only very broadband pulsed systems could be expected to produce results that differ significantly from those of CW systems." Pulse penetration depends on the dispersive nature of the dielectric medium and the Fourier coefficients (bandwidth) of the waveform.

9. R. Albanese, J. Penn, and R. Medina, "Ultrashort Pulse Response in Nonlinear Dispersive Media," in ***Ultra-Wideband, Short-Pulse Electromagnetics*****, H.L. Bertoni, L. Carin, and L.B. Felsen, Eds., Plenum, 1993, 259-265**

see also:

10. R. Albanese, J. Penn, and R. Medina, "Short-Rise-Time Microwave Pulse Propagation through Dispersive Biological Media," ***J. Opt. Soc. Amer. A*****, Vol. 6, Sept. 1989, 1441-1446**

These papers use Fourier series (of the waveform) as an analysis tool. Their calculations used a waveform with 10 carrier cycles with a carrier frequency of 1 GHz. The waveform buildup was instantaneous; the envelope was a square wave. At 1.5 m distance the rise time affected the precursor amplitude, with longer rise time reducing the amplitude. However, at this distance the main signal is very small and the precursors are also small.

11. R. Albanese, J. Blaschak, R. Medina, and J. Penn, "Ultrashort Electromagnetic Signals: Biophysical Questions, Safety Issues, and Medical Opportunities," *Aviation, Space, and Environmental Medicine*, May 1994, A116-A120

This paper shows a 1 nanosecond wideband pulse of 10 carrier cycles, carrier frequency 10 GHz. Both Sommerfeld- and Brillouin-calculated precursors appear after propagating 1 cm in pure water. Also shows precursors produced by laser (IR) radiation. Possible cellular effects of very high electric fields (hundreds or thousands of kW/m) are discussed.

12. R.A. Albanese, "An Electromagnetic Inverse Problem in Medical Science," Chapter 2 of *Invariant Imbedding and Inverse Problems*, SIAM, 1992. 30-41

Using a square-wave modulation containing 5 carrier cycles, with a 3 GHz carrier frequency, a waveform with Brillouin precursors was calculated at 10 cm. depth. With no noise, partial results were obtained, but with very weak added noise there were no results.

13. R.A. Albanese, R.L. Medina, and J.W. Penn, "Mathematics, Medicine, and Microwaves," *Inverse Problems*, Vol. 10, 1994, 995-1007

This paper shows a 1 nanosecond wideband pulse of 10 carrier cycles, carrier frequency 10 GHz. Both Sommerfeld-and Brillouin-calculated precursors appear after propagating 1 cm in pure water. Most of the paper concerns the inverse problem: given the response, find the dielectric and loss properties of the material.

14. E.L. Mokole and S.N. Samaddar, "Transmission and Reflection of Normally Incident, Pulsed Electromagnetic Plane Waves upon a Lorentz Half-Space," *J. Opt. Soc. Amer. B*, Vol. 16, May 1999, 812-831

Both reflected and transmitted fields are calculated for a semi-infinite medium; the incident waveform is a single laser pulse at IR (frequency ~2 THz). As expected, this extremely wideband waveform produces strong Sommerfeld precursors.

15. S.L. Dvorak, "Exact, Closed-Form Expressions for Transient Fields in Homogeneously Filled Waveguides," *Trans. IEEE*, Vol. MTT-42, Nov. 1984. 2164-2170

see also:

16. S.L. Dvorak and D.G. Dudley, "Propagation of Ultra-Wide-Band Electromagnetic Pulses through Dispersive Media," ***Trans. IEEE*****, Vol. EMC-37, May 1995, 192-200**

Using the Incomplete Lipschitz-Hankel Integrals (ILHI), waveform propagation in a waveguide is studied. Waveguides provide dispersion. The waveform used is a single cycle of a sine wave. Calculations show significant precursors; the waveform bandwidth is very wide. Prior experimental work of Pleshko is not mentioned, but the precursor increases in frequency as it propagates, just as measured by Pleshko.

ANNEX 5-4: RELATED PAPERS ON PRECURSORS

1. Albanese, R.A. and E.L. Bell, ***Radiofrequency Radiation and Living Tissue: Theoretical Studies*****, report SAM-TR-79-41, Dec. 1979**

This interim report models possible interactions of RF energy and cells. No distinction is made between thermal effects and pulsed effects.

2. Albanese, R.A. and E.L. Bell, "Radiofrequency Radiation and Chemical Reaction Dynamics," in ***Nonlinear Electrodynamics in Biological Systems*****, W. R. Adey and A. F. Lawrence, Eds., Plenum, 1984, 277-285**

Treated in this paper are high field strengths (100 to 2000 V/m) where the media are non-linear. There is no relevance to PAVE PAWS, where the field strengths are low, and the media are linear.

3. Albanese, R.A., R. Medina, and J. Penn, "Mathematics and Electromagnetic Theory," unpublished and undated paper, circa 2001

The statement "frequency dependence in the frequency domain implies local tissue memory in the time domain" is too general to have meaning. Certainly, frequency spectrum and time waveform are related. But this does not necessarily imply tissue memory. The entire paper is not concerned with pulses.

4. Papazoglou, T.M., "Transmission of a Transient Electromagnetic Plane Wave into a Lossy Half-Space," ***J. Appl. Phys.*****, Vol. 46, Aug. 1975, 3333-3341**

see also:

5. Dudley, D.G., T.M. Papazoglou, and R.C. White, "On the Interaction of a Transient Electromagnetic Plane Wave and a Lossy Half-Space," ***J. Appl. Phys.*****, Vol. 45, March 1974, 1171-1175**

Propagation of a single Gaussian type pulse (slow rise time and fall time) into a lossy medium is considered. One case of interest uses a lossy and dispersive medium (earth). Waveforms with and without dispersion are very much the same, probably due to the narrow bandwidth of the incident pulse. No precursors were evident.

6. Devaney, A.J., "Linearized Inverse Scattering in Attenuating Media," *Inverse Problems*, Vol. 3, 1987. 389-397

Dr. Devaney has extended this work to dispersive media, but it has not been published. The aim was to see if precursors could help in the inverse problem. Note, the inverse problem is: given the response, find the details of the scatterer or the details of the medium.

Pulse Propagation in Lossy Media

Researchers from Harvard have published papers on pulse propagation in sea water, a highly lossy medium. Single pulses and a pulse of carrier oscillations have been considered. These papers utilize a dipole as source, unlike the plane-wave source of Oughstun and colleagues. The dipole near-field greatly complicates the analysis. The papers show that the rise and fall times each produce a transient, with the carrier oscillations in between. When the rise and fall times are slow, two things occur. First, the lower frequencies in the transient allow them to decay more slowly than does the carrier. So after some distance, the carrier may be smaller than the transient waveform. These are probably not precursors in the Brillouin sense. Second, the amplitude of the transient is reduced by $f^2_{rs}/(f^2 + f^2_{rs})$, where f is the carrier frequency and f_{rs} is the principal frequency of the rise time. Thus, if the rise time equals 20 carrier cycles, the rise-time transient amplitude is reduced by 0.0025, or 52 dB.

1. King, R.W.P. and T.T. Wu, "The Propagation of a Radar Pulse in Sea Water," *J. Appl. Phys.*, Vol. 73, 15 Feb. 1993, 1581-1590

see also:

2. King, R.W.P. and T.T. Wu, Erratum: "The Propagation of a Radar Pulse in Sea Water," *J. Appl. Phys.*, Vol. 77, 1 April 1995, 3586-3687

3. King, R.W.P., "The Propagation of a Gaussian Pulse in Sea Water and Its Application to Remote Sensing," *Trans. IEEE*, Vol. GRS-31, May 1993, 595-605

4. King, R.W.P., "Propagation of a Low-Frequency Rectangular Pulse in Seawater," *Radio Science*, Vol. 28, May-June 1993, 299-307

5. Margetis, D., "Pulse Propagation in Sea Water," *J. Appl. Phys.*, Vol. 77, 1 April 1995, 2884-2888

6. Margetis, D., "Pulse Propagation in Sea Water: The Modulated Pulse," *Progress in Electromagnetics Research*, Vol. 26, 2000, 89-110

The prediction by Song and Chen that early arrival and late arrival parts of a waveform in lossy media decay algebraically (as $1/R^3$ or $1/R^5$) has been shown to be incorrect by Margetis and King:

7. Song, J. and K-M Chen, "Propagation of EM Pulses Excited by an Electric Dipole in a Conducting Medium," *Trans. IEEE*, Vol. AP-41, 1993, 1414-1421

8. Margetis, D. and R.W.P. King, "Comments on 'Propagation of EM Pulses Excited by an Electric Dipole in a Conducting Medium,'" *Trans. IEEE*, Vol. AP-43, Jan 1995, 119-120

9. Wait, J.R., "Electromagnetic Fields of Sources in Lossy Media," Chapter 24 in *Antenna Theory, Part II*, R.E. Collin and F.J. Zucker, Eds., McGraw-Hill, 1969

This paper treats single-pulse propagation in lossy (non-dispersive) media. Extensive references are provided.

ANNEX 5-5: PAPERS BY PROFESSOR OUGHSTUN AND COLLEAGUES

These papers are concerned with wideband optical (laser) short pulses in media that are lossy and dispersive. No discussions of the effects of waveform bandwidth are given. These papers are listed because they are often cited for PAVE PAWS. However the very narrow-band nature of the PAVE PAWS signal makes all of these papers irrelevant to the PAVE PAWS problem. And the laser frequency is roughly 10,000,000 times higher than that of PAVE PAWS.

1. Oughstun, K.E. and G.C. Sherman, "Propagation of Electromagnetic Pulses in a Linear Dispersive Medium with Absorption (the Lorentz Medium)," *J. Opt. Soc. of Am. B*, Vol. 5, April 1988, 817-849

2. Oughstun, K.E. and S. Shen, "Velocity of Energy Transport for a Time Harmonic Field in a Multiple-Resonance Lorentz Medium," *J. Opt. Soc. of Am. B*, Vol. 5, Nov. 1988, 2395-2398

3. Shen, S. and K.E. Oughstun, "Dispersive Pulse Propagation in a Double-Resonance Lorentz Medium," *J. Opt. Soc. of Am. B*, Vol. 6, May 1989, 948-963

4. Wyns, P., D.P. Foty, and K.E. Oughstun, "Numerical Analysis of the Precursor Fields in Linear Dispersive Pulse Propagation," *J. Opt. Soc. of Am. A*, Vol. 6, Sept. 1989, 1421-1429

5. Oughstun, K.E. and J.E.K. Laurens, "Asymptotic Description of Ultrashort Electromagnetic Pulse Propagation in a Linear, Causally Dispersive Medium," *Radio Science*, Vol. 26, Jan.-Feb. 1991, 245-258

6. Oughstun, K.E., "Pulse Propagation in a Linear, Causally Dispersive Medium," *Proc. IEEE*, Vol. 79, Oct. 1991, 1379-1390

7. Oughstun, K.E. and J.E.K. Laurens, "Asymptotic Description of Electromagnetic Pulse Propagation in a Linear Dispersive Medium," in *Ultra-Wideband, Short-Pulse Electromagnetics*, H. Bertoni et al,. Eds., Plenum Press, 1993, 223-240

8. Balictsis, C.M. and K.E. Oughstun, "Uniform Asymptotic Description of Ultrashort Gaussian Pulse Propagation in a Causal, Dispersive Dielectric," *Phys. Rev. E*, Vol. 47, 1993, 3645-3669

9. Balictsis, C.M. and K.E. Oughstun, "Uniform Asymptotic Description of Gaussian Pulse Propagation of Arbitrary Initial Pulse Width in a Linear, Causally Dispersive Medium," in *Ultra-Wideband, Short-Pulse Electromagnetics 2*, L. Carin and L.B. Felsen, Eds., Plenum Press, 1994, 273-283

10. Smith, P.D. and K.E. Oughstun, "Electromagnetic Energy Dissipation of Ultrawideband Plane Wave Pulses, in a Causal, Dispersive Dielectric," in *Ultra-Wideband, Short-Pulse Electromagnetics 2*, L. Carin and L.B. Felsen, Eds., Plenum Press, 1994, 285-295

11. Oughstun, K.E., "Transient in Chiral Media with Single Resonance Dispersion: Comments, *Jour. Opt. Soc. of Am. A*, Vol. 12, 1995, 626-628

12. Sherman, G.C. and K.E. Oughstun, "Energy-Velocity Description of Pulse Propagation in Absorbing, Dispersive Dielectrics," *J. Opt. Soc. of Am. B*, Vol. 12 Feb. 1995, 229-247

13. Oughstun, K.E., "Dynamical Structure of the Precursor Fields in Linear Dispersive Pulse Propagation in Lossy Dielectrics," in *Ultra-Wideband, Short-Pulse Electromagnetics 2*, L. Carin and L.B. Felsen, Eds., Plenum Press, 1995, 257-272

14. Oughstun, K.E., "Noninstantaneous, Finite Rise-Time Effects on the Precursor Field Formation in Linear Dispersive Pulse Propagation," *J. Opt. Soc. of Am. A*, Vol. 12, Aug. 1995, 1715-1729

15. Marozas, J.A. and K.E. Oughstun, "Electromagnetic Pulse Propagation Across a Planar Interface Separating Two Lossy, Dispersive Dielectrics," in *Ultra-Wideband, Short-Pulse Electromagnetics 3*, C. Baum, L. Carin, and A. P. Stone, Eds., Plenum Press, 1996, 217-230

16. Oughstun, K.E. and C.M. Balictsis, "Gaussian Pulse Propagation in a Dispersive, Absorbing Dielectric, "*Phys. Rev Ltrs.*, Vol. 77, 9 Sept. 1996, 2210-2213

17. Oughstun, K.E. and Hong Xiao, "Failure of the Quasimonochromatic Approximation for Ultrashort Pulse Propagation in a Dispersive, Attenuative Medium," *Phys. Rev. Ltrs.*, Vol. 78, 27 Jan. 1997, 642-645

18. Balictsis, C.M. and K.E. Oughstun, "Generalized Asymptotic Description of the Propagated Field Dynamics in Gaussian Pulse Propagation in a Linear, Causally Dispersive Medium," *Phys. Rev. Ltrs. E*, Vol. 55, Feb 1997, 1910-1921

19. Xiao, H. and K.E. Oughstun, "Hybrid Numerical-Asymptotic Code for Dispersive Pulse Propagation Calculations, *Jour. Optical Society of America A*, Vol. 15, 1998, 1256-1267

20. Solhaug, J.A., K.E. Oughstun, J.J. Stamnes, and P.D. Smith, "Uniform Asymptotic Description of the Brillouin Precursor in a Lorentz Model Dielectric," *Jour. European Opt. Soc. A., Pure and Applied Optics,* Vol. 7, 1998, 575-602

21. Oughstun, K.E., "The Angular Spectrum Representation and the Sherman Expansion of Pulsed Electromagnetic Beam Fields in Dispersive, Attenuative Media," *Jour. European Opt. Soc. A., Pure and Applied Optics,* Vol. 7, 1998, 1059-1078

22. Solhau, J.A., J.J. Stamnes, and K.E. Oughstun, "Diffraction of Electromagnetic Pulses in a Single-Resonance Lorentz Model Dielectric," *Jour. European Optical Soc. A., Pure and Applied Optics,* Vol. 7, 1998, 1079-1101

23. Smith, P.D. and K.E. Oughstun, "Electromagnetic Energy Dissipation and Propagation of an Ultrawideband Plane Wave Pulse in a Causally Dispersive Dielectric," *Radio Science*, Vol. 33, 1998, 1489-1504

24. Xiao, H. and K.E. Oughstun, "Failure of the Group-Velocity Description for Ultrawideband Pulse Propagation in a Causally Dispersive, Absorptive Dielectric," ***J. Opt. Soc. Am. B*****, Vol. 16, Oct. 1999, 1773-1785**

25. Laurens, J.E.K. and K.E. Oughstun, "Electromagnetic Impulse Response of Triply-Distilled Water," ***Ultra-Wideband, Short-Pulse Electromagnetics 4*****," Heyman et al,. Eds., Kluwer Academic/Plenum Publ., 1999, 243-264**

26. Smith, P.D. and K.E. Oughstun, "Ultrawideband Electromagnetic Pulse Propagation in Triply Distilled Water," ***Ultra-Wideband, Short-Pulse Electromagnetics 4*****," Heyman et al, Eds., Kluwer Academic/Plenum Publ., 1999, 265-276**

27. Oughstun, K.E. and H. Xiao, "Influence of Precursor Fields on Ultrashort Pulse Autocorrelation Measurements and Pulse Width Evolution," ***Optics Express*****, Vol. 8, 9 April 2001, 481-491**

28. Oughstun, K.E. and H. Xiao, "Influence of the Precursor Fields on Ultrashort Pulse Measurements," in ***Ultra-Wideband, Short-Pulse Electromagnetics 5*****, P.D. Smith and S.R. Cloude, Eds., Kluwer Academic/ Plenum Publ., 2002, 569-576**

29. Oughstun, K.E. "Asymptotic Description of Ultrawideband, Ultrashort Pulsed Electromagnetic Beam Field Propagation in a Dispersive, Attenuative Medium," in ***Ultra-Wideband, Short-Pulse Electromagnetics 5*****, P.D. Smith and S.R. Cloude, Kluwer Academic/Plenum Publi, 2002, 687-696**

6

Evidence of Biological Effects of RF Exposure Relevant to PAVE PAWS Radar System

INTRODUCTION

Biological studies to determine the plausibility of an environmental exposure having either detrimental or beneficial consequences take many forms. However, the primary goal in most of these studies is to establish a reproducible biological effect from the exposure under well-controlled conditions. Such studies utilize animal cell cultures and embryo cultures, as well as plant-growth cultures and field studies. Exposure conditions for these studies may represent those that human populations might be exposed to, or alternatively those that are expected to produce an observable biological effect in a reasonable period of time. For example, it is common research practice to use exposure levels that are much higher than those encountered in the environment to obtain a rapid and robust experimental effect. Such studies may lead to subsequent investigations of possible human health effects in animal systems or epidemiological studies. Cell-culture experiments are easy to reproduce, but the behavior of cells in culture is never exactly the same as within an organism due to the fact that cells have been isolated from their usual environment, which includes contact with a variety of other cell types, exposure to intercellular factors, and a tissue architecture. *In vitro* experiments, therefore, can be, in comparison to whole organism *in vivo* experiments, sometimes more and sometimes less likely to demonstrate a response to an environmental exposure. Cell-culture experiments are also of limited value if one wants to undertake very long-term exposure experiments (months to years). Embryonic development studies can provide greater insight into potential health effects; however, there are a limited number of established embryo models that provide well-accepted, quantifiable endpoints for study. Plant growth studies permit very long-term exposures to be studied under well-controlled conditions, but are only remotely connected to potential human health issues.

Tremendous advances in our understanding of biological systems have occurred over the 25 years since the previous review of *in vitro* studies of RF exposure effects, which was a part of the 1979 NRC review of the PAVE PAWS installation (NRC 1979). Those advances have had profound effects on our understanding of biology, greatly enabling the scientific community in its efforts to determine if and how the RF radiation generated by the PAVE PAWS system could influence biological systems. Our ability to design experiments and collect data from culture studies has been enhanced as a result of numerous technical advances, as well as significant advances in modeling the inherent complexity of biological systems, which provides a theoretical foundation for interpreting data from such experiments.

At the time of the initial NRC PAVE PAWS report (1979), cell/tissue-culture techniques were well-developed approaches but their analysis involved predominantly simple biochemical assays and single-molecule detection approaches. Protein biochemistry was also a developed field, but few specific signal-transduction pathways had been identified in cells. The techniques of molecular biology were only in the early stages of development. Today, three-dimensional tissue-culture systems are being developed to permit the study of differentiating cell systems in a native tissue architecture, and polymerase chain reaction technology permits DNA amplification from samples as small as a single cell. Differential display techniques and cDNA microarray approaches permit studies of gene expression of large numbers of genes at one time, and high performance liquid chromatography (HPLC), mass spectroscopy of "protein chips," and yeast two-hybrid systems permit studies of protein distributions, protein dynamics, and protein-protein interactions. Similarly, developments in microscopic techniques (such as atomic force, confocal fluorescent, real-time fluorescent, and micromanipulation techniques [laser capture microscopy] have significantly advanced our abilities to undertake studies of processes such as cell dynamics and intra- and inter-cellular interactions.

Of equal importance has been the development of computational approaches for modeling biological systems and responses based on concepts from complexity theory. Fuzzy logic, neural-network analysis, genetic algorithms, and clustering techniques are modern bioinformatic tools that permit the large datasets obtained by molecular biological techniques to be analyzed. These new approaches are tools that allow scientists to understand how complex systems (i.e. systems with many highly interactive components) are influenced by alterations to their environment. The computational modeling of biological processes, therefore, may help us to predict under which situations perturbations in the environment may influence biological systems. Whereas all complex systems are robustly stable to acute environmental insult, even slight perturbations in the environment can influence the self-organization of such systems if imposed for a sufficient period of time. All cells and organisms are potentially sensitive to fluctuations in the environment; however, the influence of a specific environmental perturbation on cells and organisms may not necessarily be observed in studies of a single model, but could require that a class of models be used to capture the potential set of outcomes of a given environmental perturbation. Application of these concepts to the review of biological studies of RF exposure suggests that studies that focus on the self-organization in a system (e.g., differentiation and development) and involve long-term exposures may eventually be best positioned to identify the influence of an environmental perturbation.

In order to provide a maximal representation of different cells, tissues, and "general" experimental conditions, and still remain in the realm of experimental conditions relevant to the PAVE PAWS system, this review emphasizes experimental designs from the following types of systems:

- A large ensemble of cells,
- An RF exposure similar to that generated by the PAVE PAWS system, and
- A long-term exposure to the radiation (similar to the circumstances associated with the PAVE PAWS installation).

We have also reviewed responses in cells based on stress responses that have been identified in the literature as being important for cellular recovery to acute toxicity and injury. Life on earth has evolved in an environment associated with a variety of stressors including radiation, heat, chemicals, and other toxic insults. As a consequence, cells and animals have evolved defense mechanisms to cope with these acute stresses including anti-oxidants, apoptosis (cell death to avoid replication of damaged cells), DNA repair, refolding of damaged proteins, and other mechanisms. These defense pathways work to prevent the continued expansion and use of damaged biomolecules in the cell. Much acute cellular damage can be compensated by these different damage-response pathways and so, correspondingly, these pathways represent important targets to consider in the context of detrimental biological responses to an environmental exposure. The presence of these pathways may also explain how biological responses may occur but not result in adverse health effects.

The effects of exposure to external stimuli can be categorized into:

1. Direct molecular effects
    - DNA damage
    - Membrane perturbations
    - Protein function alterations
2. Indirect molecular effects
    - Cell signal transduction

- Gene expression

3. Phenotypic effects
    - Cell apoptosis
    - Cell cycle perturbation
    - DNA repair
    - Differentiation and development
    - Carcinogenesis
    - Growth

LITERATURE REVIEW

Direct Molecular Effects

DNA Damage

DNA damage is induced by a wide variety of different stimuli including radiation, chemicals (such as chemotherapeutic drugs), and oxidative damage (including the damage that results as a normal consequence of cellular respiration). Elaborate pathways exist within cells that serve to repair damage to DNA, thereby maintaining genetic integrity and normal cellular function. Incomplete or faulty repair can lead to the production of abnormal proteins or the dysregulation of specific protein syntheses. While the overall resiliency of the cell may compensate for this dysregulation, a disruption of certain specific DNA repair processes can result in diseases such as cancer, immunodeficiency, and others. DNA damage is a normal consequence of cellular function, and it is estimated that mammalian cells encounter 10^4 DNA strand breaks/cell/day, which are usually repaired without adverse effects.

Several studies have examined possible effects of exposures to radar and cell-phone RF energies on DNA damage in mammalian cells following acute exposures (hours). The results have been contradictory and controversial (Lai and Singh 1995; McNamee and others 2002; Tice and others 2002; Bisht and others 2002; Malyapa and others 1997; Lai and Singh 1996). While Lai and Singh have detected strand breaks after RF exposures, several groups which have also used very sensitive alkaline comet assays detect no DNA break increase at pulsed RF frequencies of 1.9 GHz, 835.62 MHz, or 847.74 MHz (cell-phone frequencies) at 0.6 W/kg SAR (McNamee and others 2002; Malyapa and others 1997). Additional recent studies from Roti-Roti's laboratory have demonstrated an absence of DNA damage in brain cells, fibroblasts, and lymphoid cells following exposure to pulsed-wave microwaves and radiofrequency (Lagroye and others 2004a; Lagroye and others 2004b; Hook and others 2004). The validity of such studies depends on the sensitivity of the assay system in any particular laboratory, the types and times of exposures, and the possible sources of error. Recent studies by Dimitroglou and others (2003) have shown using single-cell gel electrophoresis higher levels of DNA strand-breaks in people who have been psychologically stressed, suggesting a role for hormonal and biochemical factors in the response, a consideration that was not previously addressed in the above reports and may be one source of error. For a review of controversial cytogenetic observations in mammalian somatic cells see Vijayalaxmi and Obe (2004).

It should be noted that the ability of an agent to cause single- or double-strand breaks (ssb or dsb) in DNA does not necessarily mean that the agent is dangerous or even capable of inducing mutations. Chemicals such as hydrogen peroxide are known to cause breaks in DNA but cause few adverse health effects. Likewise, low doses of ionizing radiation or UV light well within the limits of the permissible exposure also cause DNA strand breaks. Exposure to very low doses of ionizing radiation (1 cGy) can be detected to cause DNA strand breaks (for example, Buatti and others, 1992 find that the rate of strand breaks per cell is 1815/Gy). UV-radiation exposure, analogous to the exposure from sunlight, causes 0.07 ssb/10^{10} Da/kJ/m^2 for UVA; and 1.9 ssb/10^{10} Da/kJ/m^2 for UVB (Wenczl and others 1997).

Membrane Perturbations

Direct damage to cellular membranes has been associated with exposure to a variety of chemicals and to ionizing radiation. These stresses often lead to differences in lipid and protein composition of cellular membranes, altering the membrane fluidity, ion transport, and surface properties of the membrane.

As in the studies with DNA damage, changes in membrane fluidity or ion transport are associated with a large variety of normal states, and these changes in and of themselves are not necessarily associated with pathogenic states or with a progression toward pathogenesis. For example, chemicals known as ionophores cause a change in membrane permeability and permit ions that are usually outside the cell to rapidly influx. Nevertheless, there are very few dangerous consequences associated with exposure to ionophores, and in general they are used to mimic natural ionophoric processes that occur in mammalian cells.

Some isolated reports of changes in membranes and ion transport have been reported to be associated with exposure of cells to modulated cell-phone and radar radiation. The reported effects in the literature come largely from one laboratory, and include both changes in Ca^{++} ion flux in cells in monolayer culture and efflux from *ex vivo* tissue slices (Dutta and others, 1989; Dutta and others, 1984; Blackman and others 1985; Joines and Blackman, 1980). Subsequent studies (Cranfield and others 2001; Wolke and others 1996) have documented no or marginal effects of RF exposure on Ca^{++} flux in cell-culture systems. Taken together, these experiments provide some support for the suggestion that modulated RF exposures may be more capable of causing biological effects than unmodulated RF fields.

Clustering of membrane proteins (called membrane capping) occurs on a variety of cell types including lymphocytes and is usually a measure of membrane perturbations and membrane responses in mammalian cells. Sultan and others (1983) performed studies to determine whether or not modulated RF exposure caused changes in the ability of B-lymphocytes to cap following exposure to antigen; these studies showed no difference in the capping responses of unexposed and exposed cells in culture.

Protein Function (Folding, Protein-Protein Interactions, Translations)

Proteins are responsible for carrying out most of the cellular maintenance functions within mammalian cells. Changes in DNA are considered dangerous primarily because they lead to changes in the expression of proteins. However, cellular stress-inducing agents have been shown to affect proteins in several different ways. Changes in the ways in which proteins fold have been found predominantly following exposure to heat-shock, while changes in protein-protein interactions and in protein translation have been associated with almost all types of cellular stress. Because of the resiliency of biological systems, such damage to proteins is not usually sufficient to harm a cell unless it has occurred with a large number of proteins resulting in protein aggregation and large-scale dysfunction in a cell. In addition, cells have evolved complex systems for managing this type of damage. Chaperone systems and proteosomes are proteins that function to refold damaged proteins, clear unrepairable proteins from the cell, and disassociate aggregates created following exposure to heat-shock stress. This damage-response pathway permits heat-damaged cells to repair their proteins and resume normal function in the body. This pathway is generally induced whenever the body is experiencing fever or thermal dysregulation, although there have been a few reports in the literature of heat-like responses following exposure to ionizing radiation, UV, or other forms of stress. Numerous studies have examined the thermal effects of radar, and these effects have been well-characterized in the literature. However, such thermal effects cannot occur at the radar power densities experienced by the Cape Cod population and are not considered here.

Several recent reports have documented induction of chaperone-like heat-shock protein responses in mammalian cells following exposure to cell-phone frequencies below the thermal range. In particular, DiCarlo and others (2002), and studies by Kwee and others (1998, 2001), have documented induction of several heat-shock proteins (hsp70) in mammalian cells, as well as increased binding of heat-shock elements to their DNA-binding sequence. Replication of those findings have not been reported in the literature but are suggested by other studies showing increases in hsp70 following RF exposure (cell-phone frequencies) of fruit flys (Weisbrot and others 2003). Related studies by dePomerai and others (2000a, 2000b, 2002) have demonstrated that non-thermal microwaves can induce hsp70 in the nematode *Caenorhabditis elegans*. Similar induction of hsp70 has been observed following exposure to very low doses of ionizing radiation

(Calini and others 2003). The implications of this non-thermal induction of heat-shock proteins are not clear however they do suggest the induction of a cellular mechanism to protect proteins from degradation/denaturation. Whether this is a direct consequence of radar exposure is also not clear, but levels of induction are very low, far below the threshold of a small fever response. On the other hand, other studies have suggested that stress proteins are not induced following exposure to radiofrequency or microwave radiation (Cleary and others 1997).

INDIRECT MOLECULAR EFFECTS

Cell Signal Transduction

Extracellular signals (for growth, differentiation, apoptosis) are transmitted to the cell via a cell signal-transduction cascade that usually involves stimulation of a surface receptor, alterations in intracellular ionic concentrations, a variety of phosphorylation events, amplification of the signal, transmission of the signal into the nucleus, and alterations in gene expression. Transient changes in cell signal-transduction cascades have been associated with most cellular stress-inducing agents; these changes can lead to more permanent expressions of cellular changes such as those reflected in terminally differentiated cells.

Studies of changes in cell signal transduction following radiofrequency exposure have been limited to an examination of production of extracellular factors following exposure to RF. The most significant changes that have been reported in the literature are in the transient release of cellular growth factors and other cell signal-inducing agents that might alter cellular functions temporarily (Mausset and others 2001; George and others 2002). Other studies, however, have reported no changes in levels of such extracellular modulators as melatonin or cortisol (Radon and others 2001; Stark and others 1997). While there are numerous studies in the literature of the effects of other cell stressing agents on kinase activation or intracellular phosphorylation events, Pacini and others (2002) is the only report of changes in mitogen-activated protein kinase (MAPK) phosphorylation pathways affected by exposure to RF. Leszczynski and others (2002) reported activation of MAPK stress pathway in human endothelial cells using mobile-phone exposures.

Time-series data provide an exceptional means for characterizing cell system dynamics, and fluorescent microscopy provides a well-established technique for monitoring the activity of cell ensembles. Specifically, calcium signaling has played a major role in discussions of the influence of RF exposure on biological systems, dating to the early calcium-efflux studies of Adey and others as reported in the 1979 NRC report. Modern techniques permit the real-time monitoring of calcium activity in cell ensembles. Cranfield and others (2001) monitored calcium activity (using Fluo-3 dye) in monolayers of Jurkat E6-1 cells exposed to a 915 MHz RF waveform pulsed at 217 Hz and designed to simulate a GSM telephone signal. SAR varied from 1 to 2.1 W/kg with a weighted average of 1.5 W/kg. Mean calcium levels and calcium spike height were not affected by the RF exposure; however, a significant shift in calcium spike frequency was observed, consistent with an affect on cell ensemble activity.

Gene Expression

Changes in gene expression encompass events from activation of genes (promoter activation), through the changes in mRNA availability, to changes in protein stability/quantity, all of them finally leading to the changes in quantity of the protein product of the gene in question. Moderate changes in gene expression have been associated with mild or moderate exposures to cellular stressing agents including mild heat-shock or very low doses of ionizing radiation.

Several groups have examined changes in gene expression following exposure to RF. For the majority of genes examined for changes in expression to date, c-fos is the only transcription factor that has been shown, in multiple studies, to be modulated in response to RF exposure (Goswami and others 1999). Pacini and others (2002) reported induction of beta-transforming growth-factor and apoptosis-factor (*bax*) gene expression following RF exposure. Several studies of gene expression using comprehensive gene-chip arrays that can monitor the entire genome of expressed genes in a single experiment are underway and hold promise for resolving questions of gene expression changes associated with RF exposure.

PHENOTYPIC RESPONSES

Cell Apoptosis

Apoptosis is the process of programmed cell death induced either in natural circumstances of embryogenesis and differentiation or in states of stress following exposure to cellular damaging agents such as heat or radiation. In cell-stress responses, apoptosis can be protective because it prevents the induction of cancer cells by causing the death of cells with damaged DNA that might otherwise allow for cells bearing deleterious DNA mutations to survive.

Few studies have examined induction of apoptosis following RF exposure. As noted above, Pacini and others (2002) reported the induction of the pro-apoptotic gene *bax* following RF exposure.

Cell-Cycle Perturbations

The cell cycle not only marks the progression of a cell from single cell through its DNA replication and division into two daughter cells, but also can include stages of cellular preparation for proliferation, differentiation, or cell-cycle arrest. Cell-cycle perturbations are commonly observed following exposure to cellular stressing agents. Ionizing and UV radiations are both known to cause delays in cell-cycle progression associated with an increase in DNA repair and activation of other cell-survival functions. In addition, changes in cell-cycle time are reflective of each particular cell type and its degree of differentiation.

In studies on the effects of RF exposure, George and others (2002) demonstrated an increase in proliferative potential in exposed fibroblasts and epithelial cells. Similarly, Velizarov and others (1999) demonstrated a change in cell-proliferative capacity following RF exposure that was not related to thermal effects. Pacini and others (2002) also documented an increased proliferation of normal human skin fibroblasts in culture following exposure to RF. On the other hand, Higashikubo and others (2001) observed no changes in cell-cycle progression in two different cell lines exposed to RF (Higashikubo and others, 2001). Similarly, Stagg and others, (1997) reported no effects of a modulated RF field on cell proliferation in a glioma cell line or in primary glial cells.

These differences in observations may be due to different cell systems used, different sensitivities of assays, and different experimental conditions.

Differentiation and Development

Cellular differentiation is a critical aspect of tissue and organismal development. The process of differentiation is complex and poorly understood, yet it is known to involve changes in gene expression, protein production, cell surface-marker expression, and cell signal-transduction pathways. An undermining of this process is associated with some types of tumor progression in which normal differentiated cells are replaced by de-differentiated continuously dividing tumor cells. This process is perhaps best understood *in vitro* when differentiating agents are added to cells in culture to induce particular phenotypic changes associated with end-stage cells. A few studies have examined the effects of RF exposure on cellular differentiation.

Koldayev and Shchepin (1997) reported a study of RF effects on early embryogenesis of sea urchins. In two studies involving 450 MHz radiation for a period of 5-20 minutes, the protocol investigated fertility associated with combinations of exposed sperm and unexposed eggs, and unexposed sperm with exposed eggs. Exposure of sea urchin eggs to 100 mW/cm^2 caused a 1.2- to 1.9-fold decrease in cell fertilization. Irradiation of eggs at 200 mW/cm^2 caused a 1.8- to 2.6-fold decrease in fertilization rate, with 7-11 times the control number of abnormal zygotes. Sperm irradiation had no effect on fertilization rates. However, Saito and others (1991), in a study involving 20-day exposures of chick embryos to a 428 MHz, 5.5 mW/cm^2 incident field, showed severely delayed development according to the Hamburger-Hamilton staging. Ten replications using 10 eggs were completed, with developmental anomalies observed at SARs in the range of 8.6 mW/kg to 47.1 mW/kg.

Numerous studies have been undertaken to address the potential influence of RF exposure on teratogenesis (the production of non-heritable birth defects). In 1987, Lary and Conover (1987) completed a literature review of all teratogenesis studies extant to that date, including all studies with exposures from 300 to 3000 MHz, and below the ANSI exposure limit of 0.4 W/kg. They found no reports of teratogenic effects in the absence of organismal heating.

More recently, Heynick and Merritt (2003) reviewed work (and subsequent errata, including Saito and others 1991) on teratologic effects and developmental abnormalities from exposure to radiofrequency electromagnetic fields (RFEMF) in the range 3 kHz-300 GHz. A series of studies was conducted on beetles, birds, rodents, and nonhuman primates indicating that teratologic effects occur only from exposure levels that cause biologically detrimental increases in body temperature.

Effects of RF exposures on development in whole-animal mammalian systems are discussed in detail in Chapter 7.

DNA Repair

Normal cells have the capacity to repair damage to their genetic material as a part of their evolutionary heritage. DNA-repair pathways in mammalian cells are complex, with multiple proteins regulating the reactions. Different types of repair are associated with repair of different types of DNA damage. Faulty repair of DNA damage is associated with some human hereditary diseases such as ataxia telangiectasia or xeroderma pigmentosum. In addition, abnormalities of specific DNA-repair proteins are associated with a higher incidence of particular cancers. For example, abnormalities of the Brca1 gene, part of a DNA-repair complex, are associated with a higher incidence of breast cancer among women. Abnormalities of DNA repair would be evident in several different ways including increased accumulation of DNA damage, changes in the accumulation of particular DNA-repair proteins, and alterations in other pathways controlled by DNA-repair proteins (such as cell-cycle progression and cellular transformation). To date, there have been no direct measures of DNA-repair capacity in cells exposed to RF fields. Nevertheless, the moderate to undetectable changes observed in DNA damage, cell-cycle progression, and carcinogenesis that occur following RF exposure suggest that direct effects on DNA-repair pathways are unlikely.

Carcinogenesis is a process whereby changes in the genome of a cell lead to the progression of a normal cell into a cancerous cell. This process leads to the clonal expansion of a single cell that has uncontrolled growth and progresses to a full malignancy. Genes that have been found to be mutated in cancer cells include many of those discussed above—genes that regulate gene expression, cell-signal transduction, apoptosis, DNA repair, cellular differentiation and development, and cell-cycle progression. Neoplastic transformation is tightly associated with DNA damage and mutagenesis except in a very few cases. Reports discussed above regarding effects of RF on DNA damage relate to this issue since most DNA-damaging agents are mutagenic and also carcinogenic.

One of the most important endpoints that has been examined after RF exposures is that related to cellular transformation *in vitro* and induction of cancer in animals following exposure. Several types of *in vitro* and *in vivo* assays have been applied to the problem in an effort to discern whether or not RF exposure leads to an increased risk of carcinogenic transformation. Studying cells in culture, Roti Roti and others (2001) demonstrated that CDMA[1] radiations cause no effect on neoplastic transformation following a seven-day exposure. In related work, Cain and others (1997) reported, in a chemical tumor-promoter study using the same cell system, that modulated RF field exposure for 28 days does not lead to increased tumor promotion or progression over that of the chemical promoter alone. Detailed discussion on the effects of RF field exposure in mammals is presented in Chapter 7.

Growth in Plant Systems

A remarkable consistency has developed in a small number of studies addressing the influence of RF radiation on tree growth and fecundity, most notably in three plant studies out of the former Soviet Union in the republic of Latvia (all conducted in the vicinity of Skrunda Radio Location Station).

In the initial study, Balodis, and others (1996) reported on the growth of trees in an exposed region for the 11 years before 1971, when the radar became active, and for the 16 years following the continuous radiation exposure. Trunk samples were obtained from 50-90-year-old *Pinus sylvestris L.* trees (Balodis and others 1996) growing in 29 sampling plots and the tree heights and diameters were also measured. Radial annual increments in growth were measured to a precision of 0.01 mm. Exposure was at 156-162 MHz, with a pulse duration of 0.8 msec, interpulse interval of 41 msec, and pulse power densities as high as 375 $\mu W/cm^2$. Growth was significantly ($p = 0.001$) inhibited by exposure, with a direct linear correlation with distance from the station. Electric-field exposure levels appear to have ranged from 0.4 mV/m to 250 mV/m. Several possible sources of pollution that might have affected the ring widths were shown not to have caused the decrease in ring widths that coincided with the startup of the Skrunda radio-location system.

In a companion paper, Selga and Selga (1996) reported on Skrunda Radio Location Station EMF-induced modification of Golgi apparatus in pine needles and a switch from synthesis of predecessors of cell walls (lignins) to formation and export of resin predecessors.

[1]CPMA= Code Division Multiple Access (cellular telephone technology originally know as IS-95).

In the paper by Magone (1996), effects of Skrunda Radio Location Station EMF on the duckweed *Spirodela polyrhiza* was observed. After 55-day exposure, various morphological and developmental abnormalities were observed in 6-10 daughter plants from 10 exposed mother plants, while only 0.1 plant with abnormalities per 10 mother plants were observed in the control condition. The same daughter fronds had a shorter life-span (67 days compared to 87 days in the control) and fewer subsequent daughters (total eight compared to 10 in the control group).

In a more recent, sham-controlled study of the effect of RF exposure on tree growth, Lerchl and others (2000) obtained a similar observation. In this study, seedlings of three conifer species (*Abies alba, Abies grandis, Pinus pumila*) were exposed in a radial waveguide, in a blinded fashion, and seedling height was measured each week for a six-month period. Exposure was at 383 MHz, with a 20% duty cycle Hz, at a field intensity of 131 V/m. A significant inhibition of growth was observed.

It should be noted, however, that long-term exposures in plants can be influenced by a variety of different factors including micro-environmental changes.

SUMMARY AND CONCLUSIONS

The biological studies that most closely reflect a PAVE PAWS type of radiation exposure are the tree-growth studies. In those studies, a significant dose-dependent response was observed. In addition, some embryo development studies and a few *in vitro* studies have also suggested the possibility of ELF-modulated microwave-frequency RF exposures producing significant biological effects.

The consideration that the modulation frequency of the PAVE PAWS may play a significant role in understanding the potential for this radar to influence biological systems was also raised by the first NAS review committee in 1979. PAVE PAWS has an ELF-modulation envelope that gives rise to spectral characteristics in the 10-100 Hz frequency range. This is due to diagnostic and calibration sequences, where a pulse is removed, but many other radar systems use these sequences. Substantial research has been undertaken in this area over the last decade, with the specific intent of establishing field-intensity thresholds under well-controlled conditions. As a result of those efforts, reproducible biological effects of electric-field exposures in the ELF-frequency range have been demonstrated for induced electric-field intensities of less than 1 μ/centimeter (e.g., see the work of Rosenspire and others 2001; McLeod and Collazo 2000). Indeed, the NIEHS has concluded in a recent report that ELF electric fields below 10 μ/cm can produce reproducible biological effects (NIEHS 1998).

Responses at the cell level that are traditionally associated with carcinogenesis, such as DNA damage and mutation induction, have been observed in only one study, while several studies have documented no DNA damage associated with RF exposure. While further studies are required to clarify this issue, the evidence supporting DNA damaging events following RF exposures is weak. Other biological events that would contribute to non-DNA damage responses, and whole-organism responses, are better reported in the literature.

To clarify these conclusions, additional studies would be required. Specifically, long-term studies of complex biological responses following exposure to PAVE PAWS type radiation, as described in the U.S. Air Force Phase IV time-domain studies, would be capable of directly confirming previous reports of cell-level effects, particularly those that utilize modern large-scale data generation and analysis techniques (i.e., protein- and genomic-array-based studies). In addition, studies addressing the rectification properties of living tissue in the microwave region, most importantly, living skin, which is the predominant human organ exposed to the pulsed radiation associated with PAVE PAWS, would provide a firm foundation on which to calculate the magnitude of any induced-ELF field in the body due to exposure to ELF-modulated microwave radiation. If significant demodulation in living tissue were to be demonstrated, much of the literature regarding ELF-field interactions with tissue would be relevant to understanding the potential of PAVE PAWS exposure to influence biological systems. Similarly, due to the fact that long-term studies have demonstrated effects on plant growth from radar exposure, consideration should be given to examining plant growth around the PAVE PAWS facility.

REFERENCES

Balodis, V., G. Brumelis, K. Kalviskis, O. Nikodemus, D. Tjarve, and V. Znotina. 1996. Does the Skrunda radio location station diminish the radial growth of pine trees? Sci Total Environ 180:57-64.

Bisht, K.S., E.G. Moros, W.L. Straube, J.D. Baty, and J.L. Roti Roti. 2002. The effect of 835.62 MHz FDMA or 847.74 MHz CDMA modulated radiofrequency radiation on the induction of micronuclei in C3H 10T(1/2) cells. Radiat Res 157:506-515.

Blackman, C.F., S.G. Benane, D.E. House, and W.T. Joines. 1985. Effects of ELF (1-120 Hz) and modulated (50 Hz) RF fields on the efflux of calcium ions from brain tissue in vitro. Bioelectromagnetics 6:1-11.

Buatti, J.M., L.R. Rivero, and T.J. Jorgensen. 1992. Radiation-induced DNA single-strand breaks in freshly isolated human leukocytes. Radiat Res 132:200-206.

Cain, C.D., D.L. Thomas, and W.R. Adey. 1997. Focus formation of C3H/10T1/2 cells and exposure to a 836.55 MHz modulated radiofrequency field. Bioelectromagnetics 18:237-243.

Calini, V., C. Urani, and M. Camatini. 2003. Overexpression of HSP70 is induced by ionizing radiation in C3H 10T1/2 cells and protects from DNA damage. Toxicol In Vitro 17:561-566.

Cleary, S.F., G. Cao, L.M. Liu, P.M. Egle, and K.R. Shelton. 1997. Stress proteins are not induced in mammalian cells exposed to radiofrequency or microwave radiation. Bioelectromagnetics 18:499-505.

Cranfield, C.G., A.W. Wood, V. Anderson, and K.G. Menezes. 2001. Effects of mobile phone type signals on calcium levels within human leukaemic T-cells (Jurkat cells). Int J Radiat Biol 77:1207-1217.

de Pomerai, D., C. Daniells, H. David, J. Allan, I. Duce, M. Mutwakil, D. Thomas, P. Sewell, J. Tattersall, D. Jones, and P. Candido. 2000a. Non-thermal heat-shock response to microwaves. Nature 405:417-418.

de Pomerai, D., C. Daniells, H. David, J. Allan, I. Duce, M. Mutwakil, D. Thomas, P. Sewell, J. Tattersall, D. Jones, and P. Candido. 2000b. Microwave radiation induces a heat-shock response and enhances growth in the nematode Caenorhabditis elegans. IEEE T Microw Theory 48:2076-2081.

de Pomerai, D.I., A. Dawe, L. Djerdid, J. Allan, G. Brunt and C. Daniells. 2002. Growth and maturation of the nematode Caenorhabditis elegans following exposure to weak microwave fields. Enzyme Microb Tech 30:73-79.

Di Carlo, A., N. White, F. Guo, P. Garrett, and T. Litovitz. 2002. Chronic electromagnetic field exposure decreases HSP70 levels and lowers cytoprotection. J Cell Biochem 84:447-454.

Dimitroglou, E., M. Zafiropoulou, N. Messini-Nikolaki, S. Doudounakis, S. Tsilimigaki, and S.M. Piperakis. 2003. DNA damage in a human population affected by chronic psychogenic stress. Int J Hyg Envir Health 206:39-44.

Dutta, S.K., A. Subramoniam, B. Ghosh, and R. Parshad. 1984. Microwave radiation-induced calcium ion efflux from human neuroblastoma cells in culture. Bioelectromagnetics 5:71-78.

Dutta, S.K., B. Ghosh, and C.F. Blackman. 1989. Radiofrequency radiation-induced calcium ion efflux enhancement from human and other neuroblastoma cells in culture. Bioelectromagnetics 10:197-202.

George, F.R., R.J. Lukas, J. Moffett, and M.C. Ritz. 2002. In-vitro mechanisms of cell proliferation induction: a novel bioactive treatment for accelerating wound healing. Wounds 14:107-115.

Goswami, P.C., L.D. Albee, A.J. Parsian, J.D. Baty, E.G. Moros, W.F. Pickard, J.L. Roti Roti, and C.R. Hunt. 1999. Proto-oncogene mRNA levels and activities of multiple transcription factors in C3H 10T 1/2 murine embryonic fibroblasts exposed to 835.62 and 847.74 MHz cellular phone communication frequency radiation. Radiat Res 151:300-309.

Heynick, L.N. and J.H. Merritt. 2003. Radiofrequency fields and teratogenesis. Bioelectromagnetics Suppl 6:S174-S186.

Higashikubo, R., M. Ragouzis, E.G. Moros, W.L. Straube, and J.L. Roti Roti. 2001. Radiofrequency electromagnetic fields do not alter the cell cycle progression of C3H 10T and U87MG cells. Radiat Res 156:786-795.

Hook, G.J., P. Zhang, I. Lagroye, L. Li, R. Higashikubo, E.G. Moros, W.L. Straube, W.F. Pickard, J.D. Baty, and J.L. Roti Roti. 2004. Measurement of DNA damage and apoptosis in Molt-4 cells after in vitro exposure to radiofrequency radiation. Radiat Res 161:193-200.

Joines, W.T. and C.F. Blackman. 1980. Power density, field intensity, and carrier frequency determinants of RF-energy-induced calcium-ion efflux from brain tissue. Bioelectromagnetics 1:271-275.

Koldayev, V.M. and Y.V. Shchepin. 1997. Effects of electromagnetic radiation on embryos of sea-urchins. Bioelectrochem Bioenerg 43:161-164.

Kwee, S. and P. Raskmark. 1998. Changes in cell proliferation due to environmental non-ionizing radiation. 2. Microwave radiation. Bioelectrochem Bioenerg 44:251-255.

Kwee, S., P. Raskmark and S. Velizarov. 2001. Changes in cellular proteins due to environmental non-ionizing radiation. 1. Heat shock proteins. Electro Magnetobiol 20:165-176.

Lagroye, I., R. Anane, B.A. Wettring, E.G. Moros, W.L. Straube, M. Laregina, M. Niehoff, W.F. Pickard, J. Baty, and J.L. Roti Roti. 2004a. Measurement of DNA damage after acute exposure to pulsed-wave 2450 MHz microwaves in rat brain cells by two alkaline comet assay methods. Int J Radiat Biol 80:11-20.

Lagroye, I., G.J. Hook, B.A. Wettring, J.D. Baty, E.G. Moros, W.L. Straube, and J.L. Roti Roti. 2004b. Measurements of alkali-labile DNA damage and protein-DNA crosslinks after 2450 MHz microwave and low-dose gamma irradiation in vitro. Radiat Res 161:201-214.

Lai, H. and N.P. Singh. 1995. Acute low-intensity microwave exposure increases DNA single-strand breaks in rat brain cells. Bioelectromagnetics 16:207-210.

Lai, H. and N.P. Singh. 1996. Single- and double-strand DNA breaks in rat brain cells after acute exposure to radiofrequency electromagnetic radiation. Int J Radiat Biol 69:513-521.

Lary, J.M. and D.L. Conover. 1987. Teratogenic effects of radiofrequency radiation. IEEE Eng Med Biol Mag. march 1987, 42.

Lerchl, D., A. Lerchl, P. Hantsch, A. Bitz, J. Streckert, V. Hansen. 2000. Studies on the effects of radio-frequency fields on conifers. Trans Bioelectromagnetics Soc 22:160.

Leszczynski, D., S. Joenvaara, J. Reivinen, and R. Kuokka. 2002. Non-thermal activation of the hsp27/p38MAPK stress pathway by mobile phone radiation in human endothelial cells: molecular mechanism for cancer- and blood-brain barrier-related effects. Differentiation 70:120-129.

Magone, I. 1996. The effect of electromagnetic radiation from the Skrunda Radio Location Station on Spirodela polyrhiza (L) Schleiden cultures. Sci Total Environ 180:75-80.

Malyapa, R.S., E.W. Ahern, W.L. Straube, E.G. Moros, W.F. Pickard, and J.L. Roti Roti. 1997. Measurement of DNA damage after exposure to 2450 MHz electromagnetic radiation. Radiat Res 148:608-617.

Mausset, A.L., R. de Seze, F. Montpeyroux, and A. Privat. 2001. Effects of radiofrequency exposure on the GABAergic system in the rat cerebellum: clues from semi-quantitative immunohistochemistry. Brain Res 912:33-46.

McLeod, K.J. and L. Collazo. 2000. Suppression of a differentiation response in MC-3T3-E1 osteoblast-like cells by sustained, low-level, 30 Hz magnetic field exposure. Radiat Res 153:706-714.

McNamee, J.P., P.V. Bellier, G.B. Gajda, B.F. Lavallee, E.P. Lemay, L. Marro, and A. Thansandote. 2002. DNA damage in human leukocytes after acute in vitro exposure to a 1.9 GHz pulse-modulated radiofrequency field. Radiat Res 158:534-537.

NIEHS (National Institute of Environmental Health Sciences). 1998. Portier, C.J. and M.S. Wolfe (eds). Assessment of health effects from exposure to power-line frequency electric and magnetic fields. Working Group Report. NIH Publication 98-3981.

NRC (National Research Council). 1979. Analysis of the Exposure Levels and Potential Biologic Effects of the PAVE PAWS Radar System. Washington, DC: National Academy Press.

Pacini, S., M. Ruggiero, I. Sardi, S. Aterini, F. Gulisano, and M. Gulisano. 2002. Exposure to global system for mobile communication (GSM) cellular phone radiofrequency alters gene expression, proliferation, and morphology of human skin fibroblasts. Oncol Res 13:19-24.

Radon, K., D. Parera, D.M. Rose, D. Jung, and L. Vollrath. 2001. No effects of pulsed radio frequency electromagnetic fields on melatonin, cortisol, and selected markers of the immune system in man. Bioelectromagnetics 22:280-287.

Rosenspire, A.J., A.L. Kindzelskii, and H.R. Petty. 2001. Pulsed DC electric fields couple to natural NAD(P)H oscillations in HT-1080 fibrosarcoma cells. J Cell Sci 114:1515-1520.

Roti Roti, J.L., R.S. Malyapa, K.S. Bisht, E.W. Ahern, E.G. Moros, W.F. Pickard, and W.L. Straube. 2001. Neoplastic transformation in C3H 10T(1/2) cells after exposure to 835.62 MHz FDMA and 847.74 MHz CDMA radiations. Radiat Res 155:239-247.

Saito, K., K. Suzuki, and S. Motoyoshi. 1991. Lethal and teratogenic effects of long-term low-intensity radio frequency radiation at 428 MHz on developing chick embryo. Teratology 43:609-614.

Salford, L.G., A.E. Brun, J.L. Eberhardt, L. Malmgren, and B.R. Persson. 2003. Nerve cell damage in mammalian brain after exposure to microwaves from GSM mobile phones. Environ Health Persp 111:881-883; discussion A408.

Selga, T. and M. Selga. 1996. Response of Pinus sylvestris L needles to electromagnetic fields. Cytological and ultrastructual aspects. Sci Total Environ 180:65-73.

Stagg, R.B., W.J. Thomas, R.A. Jones, and W.R. Adey. 1997. DNA synthesis and cell proliferation in C6 glioma and primary glial cells exposed to a 836.55 MHz modulated radiofrequency field. Bioelectromagnetics 18:230-236.

Stark, K.D., T. Krebs, E. Altpeter, B. Manz, C. Griot, and T. Abelin. 1997. Absence of chronic effect of exposure to short-wave radio broadcast signal on salivary melatonin concentrations in dairy cattle. J Pineal Res 22:171-176.

Sultan, M.F., C.A. Cain, and W.A. Tompkins. 1983. Immunological effects of amplitude-modulated radio frequency radiation: B lymphocyte capping. Bioelectromagnetics 4:157-165.

Tice, R.R., G.G. Hook, M. Donner, D.I. McRee, and A.W. Guy. 2002. Genotoxicity of radiofrequency signals. I. Investigation of DNA damage and micronuclei induction in cultured human blood cells. Bioelectromagnetics 23:113-126.

Velizarov, S., P. Raskmark, and S. Kwee. 1999. The effects of radiofrequency fields on cell proliferation are non-thermal. Bioelectrochem Bioenerg 48:177-180.

Vijayalaxmi, Obe. G. 2004. Controversial cytogenetic observations in mammalian somatic cells exposed to radiofrequency radiation. Radiat Res 162:481-496.

Weisbrot, D., H. Lin, L. Ye, M. Blank, and R. Goodman. 2003. Effects of mobile phone radiation on reproduction and development in Drosophila melanogaster. J Cell Biochem 89:48-55.

Wenczl, E., S. Pool, A.J. Timmerman, G.P. van der Schans, L. Roza, and A.A. Schothorst. 1997. Physiological doses of ultraviolet irradiation induce DNA strand breaks in cultured human melanocytes, as detected by means of an immunochemical assay. Photochem Photobiol 66:826-830.

Wolke, S., U. Neibig, R. Elsner, F. Gollnick, and R. Meyer. 1996. Calcium homeostasis of isolated heart muscle cells exposed to pulsed high-frequency electromagnetic fields. Bioelectromagnetics 17:144-153.

7

Animal and Human Studies Addressing Health Effects

INTRODUCTION

Studies to evaluate the potential of RF fields to impact biological systems have been conducted in both humans and animals. Although the interaction of RF fields with humans is of prime importance and concern, the number of human laboratory studies is quite small compared to the literature available on animal studies. Moreover, many areas of biologic investigation are more efficiently and appropriately conducted using various animal species. Animal studies provide an integrated system that can be used in studies where experimental variables can be controlled, specific hypotheses can be explored, and exposure can be precisely assessed. Given the uncertainty and the relatively low power of RF epidemiologic studies to ascertain the relationship between RF exposure and possible adverse health effects, and the small number of human laboratory studies, investigations in animals are especially important in the evaluation of RF for potential adverse effects.

There are, however, limitations to animal studies for risk-assessment purposes that one must bear in mind. Extrapolation of experimental results from laboratory animals to humans remains somewhat tentative, both because of important biologic differences between humans and animals and because the mechanisms (and hence the biologically relevant exposure parameters) by which the effects may arise are often unknown. Furthermore, studies using laboratory animals are typically carried out at exposure levels much higher than found in the environment. There are questions as to the applicability of these laboratory study conditions to the low RF levels experienced by humans exposed to the PAVE PAWS signal.

It should be noted that there is a considerable literature investigating the thermal effects of RF exposure in animals. Much of this literature is focused on 2450 MHz experimentation (outside the frequency range of interest for the PAVE PAWS radar) although other frequencies are also occasionally investigated. Because thermal responses are clearly outside the range of expected biologic responses to the PAVE PAWS exposures, this report does not consider in detail work performed at exposure levels that are designed to produce elevated temperatures in exposed animals. Reference to exposures at 2450 MHz is made occasionally where studies at the more relevant frequencies are lacking; however, the 2450 MHz work is not reviewed comprehensively.

CANCER—HUMAN STUDIES

There are no laboratory studies in humans with exposures to RF energy and cancer as the direct endpoint of investigation. Such studies would clearly cross ethical boundaries and would not be acceptible. Futhermore, RF fields lack sufficient energy to disrupt chemical bonds so there is little theoretical basis for suspecting that such fields would cause mutations or other genotoxic effects. There are a few human studies investigating possible indirect and nongenotoxic effects relevant to cancer that will be discussed in the section on immunological and endocrine function studies.

CANCER—STUDIES IN ANIMAL MODELS

Several different approaches and animal models have been used in laboratory animal cancer studies. The selection of a specific model depends largely upon the hypothesis chosen to evaluate a particular underlying mechanism. For example, if one desires to test RF for its potential to be a complete carcinogen (an agent that by its application alone can cause cancer), animals are exposed to RF over a long period of time, usually 1 1/2 to 2 years. The extended period of exposure is necessary to ensure adequate time for slowly developing cancers to be manifest. During the exposure period, whatever the length, it is important to keep exposures to other possibly confounding agents to a minimum. In this regimen, the animals are usually observed during the major portion of their lifetime and the occurrence of tumors, in number, type, and time of development, are the critical endpoints. This type of study should include several dose groups and requires a relatively large number of animals, particularly if the natural incidence of a tumor type is low. As one would surmise, studies evaluating complete carcinogenicity are expensive due to both the length of time and the number of animals involved.

Because carcinogenesis is a multi-step process, another experimental approach is to assume the agent of interest (RF) acts as either an initiator or a promoter where a two-phase protocol is required for testing. “Initiation” is defined as a genotoxic event where the carcinogen interacts with the DNA and introduces genetic changes that may later result in malignancy. "Promotion" is operationally defined, where the promoting agent is applied subsequent to initiation and generally over a protracted period of time. Promotion is associated with a number of subcellular events that are generally non-genotoxic and is responsible for the conversion of initiated cells to cancerous cells. To evaluate RF as an initiator, one high dose of RF would be given, followed by repeated exposure to a known promoter (e.g., 12-O-tetradecanoylphorbol-13-acetate, TPA) over a long term. If RF were to be investigated for possible promotional effects, the animals would be treated with a known initiator (e.g., 7,12-dimethyl benz[a]anthracene, DMBA), and subsequently exposed to RF over a long term (months). These initiation/promotion approaches have the advantage of using fewer animals over a shorter period of time resulting in less cost. However, a given model is usually limited to evaluating a specific type of cancer. Because current knowledge on possible biological mechanisms of the RF exposures is limited, other than thermal effects of high doses, the applicability of these studies to cancer development in humans exposed to RF may also be limited.

Long-Term Animal Bioassays

Long-term animal bioassays, often conducted in two species (usually rats and mice), and in both male and female animals for two years, provide a reasonable surrogate for human-lifetime exposure. A relatively small number of long-term animal bioassays have been performed exposing rats and mice to RF signals between 10 and 2000 MHz. Almost all of the studies performed at non-thermal levels have indicated no pathological or carcinogenic effects. This includes studies with a focus on brain cancer at 836 and 860 MHz (1.1 to 1.6 W/kg, Adey and others 1999, 2000; ~1 W/kg, Zook and Simmens 2001, respectively), as well as complete histopathology in lifespan and hematology studies at 835/847, 800, and 2450 MHz (1.3 W/kg, LaRegina and others 2003; up to 12.9 W/kg, Spalding and others 1971; 0.3 W/kg, Frei and others 1998; respectively). Although some pathological effects have been reported at thermal levels (Roberts and Michaelson 1983; Prausnitz and Susskind 1962), the only report of an increased tumor incidence with long-term RF exposure at non-thermal levels was by Chou and others (1992). They reported a small increase in overall tumor occurrence in rats exposed for 24 months to 2450 MHz (SAR of 0.15-0.4 W/kg). There was no effect in the Chou and others study on a number of other parameters including metabolism, immune function, hematology, serum chemistry, thyroxin levels, protein parameters, growth, or open-field behavior. It should be noted that the one bioassay cancer study investigating a frequency (435 MHz) most closely related to the PAVE PAWS frequency also found no overall increase in cancer in exposed animals (1.0 mW/cm^2; Toler and others 1997). Some organs (specifically the adrenal glands) did show slight trends toward increased cancer in RF exposed animals; however, the number of tumors was small and no statistically significant differences could be determined between the exposed and control groups.

Radiation- or Chemically Initiated and Transgenic Animal Studies

Similar to long-term animal bioassays, studies in which tumors have been initiated by means other than RF have been mostly negative. Many different initiation models have been used in these studies in which rodents have been exposed to radiofrequencies between 800 and 1500 MHz. The experimental models used include: brain tumors initiated in rats with ethyl nitrosourea (836 and 860 MHz, approximately 1 W/kg) (Adey and others 1999, 2000; Zook and others 2001, respectively); benz(a)pyrene initiated sarcomas in rats (900 MHz, 0.075 and 0.27 W/kg) (Chagnaud and others 1999); dimethyl-benzanthracine-initiated rat mammary tumors (900 MHz, SARs from 0.017 to 0.07 W/Kg) (Bartsch and others 2002; Anane and others 2003); diethyl nitrosamine-induced hepatomas in rats (929 and 1500 MHz) (Imaida and others 1998); and radiation-induced mouse lymphomas (902 MHz, 0.35 W/kg) (Heikkinen and others 2001). In all of these cited cancer studies, no adverse effects of RF exposure were noted.

In addition to chemicals and radiation, genetically initiated animal models (transgenic mice) have also been studied in RF carcinogenicity testing. No effects of RF exposure on mutagenicity or tumor development were found using pKZ-1 transgenic mice (900 MHz, 4 W/kg) (Sykes and others 2001). Another study by Repacholi and others (1997) in Pim-1 mice reported an association between long-term RF exposure [900 MHz, 0.13-1.4 W/kg] and mortality from a certain subtype of lymphoma. A subsequent study performed at multiple dose levels with more uniform and more fully characterized exposure (900 MHz, multiple levels to 4.0 W/Kg) did not confirm the positive effects reported in the original study (Utteridge and others 2002).

Tumor-Cell Injection Studies

A few studies of tumor progression, using non-thermal RF exposure levels, have been conducted by injecting tumor cells into mice and determining growth rate, survival, and metastatic progression. Although increased survival of the host, as well as inconsistent evidence of either augmentation or suppression of immune function has been reported in response to thermal levels of RF exposure, no such effects were observed in studies using lower levels of exposure (Salford and others 1993, 915 MHz, up to 8.3 W/kg; Higashikubo and others 1999, 836 and 847 MHz, 0.75 W/kg).

Summary of Animal Cancer Studies

Most animal bioassay studies have not demonstrated increased cancer risk resulting from long- or short-term RF exposure at non-thermal levels. In the very few studies at thermal levels within the frequency range of interest (10-2000 MHz), only inconsistent evidence of exposure effects have been reported, and those have not been confirmed in similar or replicate studies.

HUMAN BEHAVIORAL STUDIES

Most of the RF behavioral studies in humans have focused on frequencies associated with cellular telephony (800/900 to 1800 MHz). Areas of investigation have included hypersensitivity to exposure, sleep, memory, attention behaviors, or other cognitive functions.

A wide range of subjective health responses have been attributed to RF exposures (see review, Sandstrom and others 2001), including headaches, fatigue, dizziness, and nausea. The number of reports, however, from studies actually conducted under controlled research conditions, is quite small. In those blind studies (where exposure or non-exposure status was not known by the researchers or subjects during experimentation), Koivisto and others (2001, Hietanen and others (2002), Zwamborn and others (2003), found no effects of RF exposure.

The studies examining sleep patterns in people exposed to RF fields have presented mixed results with some effects reported, although the positive responses remain relatively ill-defined. An initial report by Mann and Roschke (1996) suggested that RF exposure (900 MHz) resulted in decreased latency to sleep onset and a reduction in rapid eye movement (REM) sleep. These observations were not replicated in further studies (Wagner and others 1998, 2000). Borbely and others (1999) and Huber and others (2003) reported changes in the spectral power in non-REM sleep but REM sleep and onset latency were unaffected. A reduction in the percentage of slow-wave sleep was observed by Lebedeva and others (2001) when subjects were exposed throughout the night. In humans exposed to a 900 MHz at 1 W/kg SAR, Huber and others (2000, 2002) observed changes in the spectral power of EEG patterns in the initial phases of sleep, but when the exposure was to a continuous-wave signal, no significant effects occurred. Two other recent studies (Mann and Roschke, 2004) found no changes in sleep patterns and no evidence of sleep disturbances due to RF exposure.

Cognitive function has been the focus of a number of recent studies evaluating performance and RF exposure. Preece and others (1999) observed decreased choice reaction time in people exposed to analog 902 MHz fields. However, digital signals did not produce such changes. Nor were there effects from exposure in simple reaction times or in spatial memory. The choice reaction time changes were ascribed to possible thermal impacts of the signal. In another study at 902

MHz, Kovisto and others (2000) reported decreased response times in simple reaction and vigilance tasks with localized heating in the brain as the possible explanation for the effects. Using an improved experimental design, the same research group was unable to confirm the inital findings and reported no changes in the reaction times or error rates with exposure (Haarala and others 2003). There have been a few additional studies examining different cognitive functions and memory. Many of these studies show improved performance on recognition memory task (Lass and others 2002), improvement on cognitive tasks (Edelstyn and Oldershaw 2002), a field-dependent improvement in memory (males only) at 1800 MHz (Smythe and Costall 2003), and a facilitating effect on attention in mobile phone users (Lee and others, 2001).

ANIMAL BEHAVIOR

Alteration of behavior in animals has provided the foundation for human RF-exposure guidelines for the past two decades (ANSI C95.1-1982; NCRP, 1986; IEEE/ANSI C95.1-1992). However, nearly all of the reported studies, including those showing behavioral responses in exposed animals, have been conducted in an RF-intensity range that would be expected to produce thermal sensations and/or heating of tissue (e.g., Brown and others 1994; Akyel and others 1991; D'Andrea and others 2003).

Acute thermal responses in animals can include perception, aversion, work perturbation or work stoppage, decreased endurance, and even convulsions and death. Behavioral effects of RF in the non-thermal range, however, are more difficult to identify. Studies usually conducted in mice or rats and using non-thermal levels of RF exposure (and even some using thermal levels of exposure), have generally reported no effects on various aspects of behavior, including operant behavior (1.25 GHz at less than 7.6 W/kg, Akyel and others 1991; 1.3 GHz at less than 3 W/kg, Lebovitz 1981 and Lebovitz and Seaman 1983; 900 MHz at 17.5 to 75 mW/kg, Bornhausen and Scheingraber 2000); cognitive behavior (900 and 1800 MHz at 0.5 W/kg, Sienkowicz and others 2000); and performance and activity changes in rats exposed in utero (915 MHZ at 3.6 W/kg, Jensh and others 1982a,b). An additional study examined chick behavior (450 MHz at 5 mW/cm^2, Sagan and Medici 1979). One report does suggest behavioral changes with apparent non-thermal acute RF exposures, including reduced aggressive behavior in rats (1.3 GHz at 0.65 mW/cm^2, Frey and Spector 1986). Additionally, performance in spatial-navigation tests has been examined in a number of studies using radial-arm or water mazes (at 900 and 1800 MHz, 0.5 W/kg, Sienkiewicz and others 2000, and at 900 MHz, up to 3.5 W/kg, Dubreuil and others 2002), with no indication of adverse effects. There has been a report of reduced performance in a radial-arm maze at 2450 MHz by Lai and others (1994); however, Cobb and others (2004) were unable to demonstrate similar results in rats, also exposed at 2450 MHz, and other studies have also been unable to confirm Lai's results

(Cosquer and others 2004; Cobb and others 2004; Cassel and others 2004). Yamaguchi and others (2003) reported no effects in a T-maze performance study at 1.4 GHz until the exposure reached clearly thermal levels (25 W/kg).

With chronic low-level RF exposures, reports on behavioral effects have been generally negative (D'Andrea and others 1980), although positive reports at near-thermal levels (2.7 W/kg whole-body average) have been reported (Mitchell and others 1977, 1981, 1988) at 2450 MHz.

Summary of Behavioral Studies

Results from laboratory studies in humans have indicated subtle and transient effects; however, the health implications remain unclear. There is some evidence that acute exposure may result in minor facilitation effects on attention functions and decreases in some specific reaction times to stimuli. Effects on sleep have been reported but remain mixed and ill-defined. The available data are too sparse to determine if subjective responses can be caused by RF exposure, although the strongest studies indicate no effects on a range of endpoints.

Disruption of complex behavioral performance in several animal species, under diverse exposure conditions, has been used as a basis for setting human exposure guidelines since 1982. The threshold SAR selected to establish the standard was chosen at 4 W/kg, a level based upon thermal effects and often (but not always) accompanied by an increase in core body temperature of ~1.0°C. Alteration of an assortment of other behaviors, both learned and unlearned, can also occur in animals at SARs between 1and 4 W/kg, subject to the frequency of the signal and the size of the animal. It appears that the behavioral changes due to RF exposure at these levels are reversible, and no consistent evidence exists for long-term, permanent effects. Extrapolation of animal data to humans has been useful in setting exposure standards. However, human ability to discriminate and cognitively act upon perception of intense RF fields is generally superior to that ability in animals and therefore animal data may tend to underestimate threshold levels for safety.

OTHER PHYSIOLOGICAL STUDIES

EEG and Brain Electrical Activity

Humans exposed to mobile-phone RF fields have generally not shown effects of exposure on the spontaneous, awake electroencephalogram (EEG) (Hietanen and others 2000; Roschke and Mann 1997). However, possible changes have been observed when exposure was given during the performance of memory tasks or under other more demanding paradigms (Freude and others 1998, 2000; Eulitz and others 1998. Krause and others reported changes in EEGs of humans exposed to 902 MHz for both auditory tasks (Krause and others

2000a) and visual memory tasks (Krause and others 2000b). Additionally, Croft and others (2002) determined the resting EEG parameters to be changed during an auditory discrimination task. Several studies evaluating sleep parameters have examined EEG in human subjects as a way to assess RF impacts on sleep. These studies have been discussed above in the section on behavior and RF exposure in humans.

Studies in a variety of animals, including rats, rabbits, cats, and monkeys have shown various changes in EEG measurements from the brain (rabbits at 30 MHz, 0.5-2 kV/m, Takashima and others 1979; cats at 147 MHz, unspecified SAR, Bawin and others 1973; rats at 945 MHz, unspecified SAR, Vorbyov and others 1997; rats at 900 MHz, 1.3 W/kg, Thuroczy and others 1994); however the types of changes are not consistent across studies nor have they generally been independently and systematically confirmed.

Effects on Blood Pressure/Heart Rate

A stimulation study in humans by Braune and others (1998) initially reported increases in heart rate (HR) and blood pressure (BP). However, these effects were not replicated in the same laboratory (Braune and others 2002), and were also not confirmed by an additional independent human study (Tahvanainen and others 2004). Animal studies have reported effects of RF exposure on BP and HR and other cardiac functions (e.g., Lu and others 1992); however, these have all been conducted at exposure levels in which thermal increases would be expected in the tissue.

Blood Brain Barrier Studies

Using RF exposures of 2450 MHz, Frey and others (1975) initially reported that at approximately 1 W/kg an increase in blood brain barrier (BBB) permeability was observed in rats. Using a 1.3 GHz RF signal, Oscar and Hawkins (1977) reported increased BBB permeability at 0.4 W/kg (CW) and 0.1 W/kg (PW). Preston and others (1979) suggested that the changes observed by Oscar and Hawkins may have been due to variations in blood flow, so Oscar and others (1981) subsequently determined that increased local brain blood flow did occur following RF exposure. In following up this finding, Oscar and co-workers used a technique to measure BBB permeability that is insensitive to blood-flow change, and reported no effect of RF radiation on BBB (Gruenau and others 1982). Thus the effect originally reported by Oscar and Hawkins (1977) was probably an artifact. Using techniques similar to those of Oscar and Hawkins, at 2450 MHz, Preston and others (1979) and Preston and Prefontaine (1980) reported no effect of RF exposure on BBB permeability at whole body SARs (0.02-6 W/kg) or at SARs in the head (0.08-1.8 W/kg). Additional efforts to establish the occurrence of the BBB effects observed by Oscar and Hawkins (1977) and Frey and others

(1975) have been unsuccessful (Ward and others 1982; Ward and Ali 1985; Merritt and others 1978).

Pigs have been exposed repeatedly to 452 MHz fields intermittently for 8 h/d for 90 days. The BBB showed no leakage in exposed animals, nor did neurohistological and enzyme-histochemical preparations show any evidence of damage to nervous tissue in the brain (Sutton and others 1982). Further animal studies in mice also have demonstrated no BBB permeability changes with either a one-hour exposure at 4 W/kg (whole body) or after a lifetime of exposure at SARs ranging from 0.25, 0.5, 1.0, and 4.0 W/kg (whole body) (Finnie and others 2001, 2002).

At 2450 MHz, Sutton and Carroll (1979) observed that BBB permeability was increased in rats when the temperature in the brain was 40° C or more. Furthermore, when the core body temperature of the rat was kept at 30° C during exposure of the head, the exposure time had to be lengthened to produce any disruptive effects on the BBB. These observations suggest that RF-induced hyperthermia may indeed be the cause of BBB disruption during exposure. Merritt and others (1978) also showed that BBB permeation in rats was impacted by providing either hot air or RF radiation to heat the animals to 40°C and concluded that hyperthermia was the causative factor, not RF energy per se. Williams and colleagues (1984a-d) report on a series of experiments in which they conclude that RF exposure (at 2450 MHz) produced BBB effects that result from temperature-dependent changes and not as a direct result of the RF energy. Fritze and others (1997) also found BBB permeability changes in rats consistent with thermal effects. A number of other papers have also demonstrated changes in BBB permeability resulting from thermal effects of RF exposure (Lin and Lin 1980, 1982; Goldman and others 1984; Neilly and Lin 1986; Moriyama and others 1991; Ohmoto and others 1996).

There are a few papers that report BBB-permeation effects in animals exposed to RF fields below those considered to be "thermal." Persson and others (1997) observed an increase in BBB permeability by about three-fold in animals exposed to CW 915 MHz radiation. However, the changes did not vary with SAR from 0.02 to 8.3 W/kg. Results using modulated RF exposure were not SAR dependent either, with the lowest SARs (0.0004-0.008 W/kg) demonstrating the greatest permeability changes, and at the highest SARs (1.7-8.3 W/kg) no modulated frequency was effective in increasing BBB permeation. The 1997 paper by Persson and others appears to include data from previous studies in their laboratory (Salford and others 1993, 1994; Persson and others 1992) and a recent paper (Salford and others 2003) from this research laboratory describes effects of 915 MHz RF on the BBB in rats exposed to very low SARs (< 0.2 W/kg). No exposure-response relationship was found in work performed in another laboratory (Chang and others 1982) in which only one of six RF exposure levels (at 1 GHz) affected BBB permeability in dogs.

IMMUNE- AND ENDOCRINE-FUNCTION STUDIES

There are very few studies in humans examining endocrine status during or following exposure to RF fields. Mann and others (1998) observed no changes in serum melatonin, growth hormone, or luteinizing hormone levels during exposure of volunteers to 900 MHz, although cortisol production showed a small transient increase. No effect on urinary levels of 6-hydroxymelatonin sulfate was reported in humans exposed to 900 MHz fields (Bortkiewicz and others 2002). In addition, Radon and others (2001) observed no changes in melatonin, cortisol, or markers of immune function when humans were exposed for 4 hours to 900 MHz fields.

In animals, most of the studies investigating immune and endocrine function have been performed at 2450 MHz, with only a few in the 10 MHz to 2000 MHz range. A few of those laboratory studies report alterations in various hormones (Abhold and others 1981) and neurotransmitters (Inaba and others 1992; Mausset and others 2001) at 900 MHz in animals exposed to low intensities (non-thermal levels) of RF. There are reported increases as well as decreases in immune-cell subpopulations, exposed at 900 MHz (Dasdag 2000), and decreased levels of immunoglobin titer and cellular-immunity function depending upon modulation frequency of the RF signal used for exposure (450 MHz, Lyle and others 1983).

Most of the low-level exposure studies indicate no significant changes in hormone levels or activity (900 MHz, Vollrath and others 1997, and Heikkanen and Juutilainen 1999). In addition, a number of studies have been reported in which RF exposure, at levels insufficient to cause increased temperatures in tissue, does not produce observable changes in immune cell function, differentiation, mitogenic activity, or other hematological parameters (Djordjevich and others 1977; 100 MHz, Smialowicz and others, 1981a,b; 900 MHz, Chagnaud and Veyret 1999).

A notable exception to these studies is the work of Toler and others (1988) who studied the changes in blood-borne factors during six months of exposure to a 435 MHz radiation at 0.3 W/kg. While most factors were not observed to undergo significant change, dopamine levels were found to drop almost immediately at the start of the exposure period, and remained depressed throughout the six-month exposure period. Dopamine levels at the end of the study period were only one-half those in the sham-exposed animals. Although not yet replicated, the results of this study are important to consider due to its size, the magnitude and duration of altered dopamine response, and because the exposure frequency (435 MHz) is the center frequency of the PAVE PAWS radar.

Those studies in which effects have been observed in immune-system parameters or endocrine function are predominantly at exposure levels at 2450 MHz that are clearly in the thermal RF range (Gildersleeve and others 1988; Lu and others 1985, 1986, 1987; Michaelson and others 1961). Using thermal levels of RF exposure, one study found no effect on autoimmune response (Anane and others 2003); however, many studies observed either increased or decreased immune-cell function (Bogolyubov and others 1987, 1988; Liburdy 1977, 1979, 1980; Takashima and Asakura 1983; Smialowicz and others 1981a,b, 1982a, 1982b) as well as the induction of stress markers (Cleary and others 1980). The reported effects appear to be similar to the effects of non-RF heating.

Summary for Other Physiological Parameters

The originally reported effects of low-level RF exposure on the BBB have not been confirmed. However, many investigators have produced results that indicate changes in the permeability of the BBB when a significant increase in temperature occurs as a result of absorption of RF energy. The thermal effects have been demonstrated through a range of endpoints, including uptake of radiotracers, dyes, and proteins. Some studies have even shown uptake of virus particles and drugs to be influenced by RF-produced thermal increases. Based on modeling studies, localized exposure of the head at 1.6 W/kg will produce a 0.1°C increase in brain temperature, an increase that is small in comparison to the temperature increases associated with changes in BBB permeability described above. The reports of permeability changes in the BBB at SARs <4 W/kg generally are not useful in arriving at exposure guidelines since the effects at these low levels have not been confirmed and no dose-response relationship has been established.

RF studies investigating hematologic, immunologic, and endocrinologic endpoints in animals exposed to RF have produced both positive and negative results. However, most of the studies in which effects were observed have been performed at intensities of RF exposure expected to result in increased temperatures. In the few studies that have reported effects at non-thermal exposure levels, the observations are inconsistent across studies and not in general agreement with the larger body of evidence pointing toward non-effects at these levels. Certainly, there is a lack of evidence of RF-induced effects that can be directly related to adverse health responses.

TERATOLOGY, REPRODUCTION, AND DEVELOPMENT

Teratogenicity

During pregnancy, heat stress in animals where maternal body temperatures are raised to thresholds of 40.5-42.0°C has been demonstrated to increase the incidence of birth defects (Boreham and others 2002; Bennet and others 1990). Similarly, RF exposure of pregnant rodents, sufficient to increase maternal core body temperature (whole-body average SARs of > 9 W/kg), has also been shown to be teratogenic (27 MHz, Lary and others 1982, 1983, 1986; 970 MHz, Berman and others 1992; 915 MHz, Jensh and others 1982a, 1997; Guillet and Michaelson 1977). Exposure to RF at ~3 W/kg at 2450 or 100 MHz has been reported to cause a decrease in Purkinje cells in the cerebellum of neonatal rats (Albert and others 1981a) but not squirrel monkeys (only at 2450 MHz, Albert and others 1981b). At reduced exposure levels of 2450 MHz RF, lower than those causing malformations but still thermal in nature (~4-5 W/kg), fetal mass appears to be diminished (Marcickiewicz and others 1986).

Lower levels of RF exposure that produce no significant thermal elevation in tissue have not been associated with teratogenesis (Schmidt and others 1984). Continuous exposure of rats during gestation at 0.4 W/kg (2450 MHz) produced no effect on brain development, brain weight, or DNA, RNA, and protein content (Merritt and others, 1984). A long-term exposure study in squirrel monkeys, also conducted at 2450 MHz, with whole-body exposures of up to 3.4 W/kg, found no effect in a broad array of endpoints including birth defects, development, behavior, EEG, biochemistry, and hematology (Kaplan and others 1982). In contrast, after exposure of rats to 27 MHz, Tofani and colleagues (1986) have reported teratogenic effects in the offspring after maternal exposure at whole-body average SARs as low as 0.0001 W/kg. Clearly, this study, which has not been replicated independently, is inconsistent with the majority of laboratory evidence reporting teratogenic effects of RF exposure only when associated with increases in maternal temperature. Low-level exposures at 428 MHz were reported to cause a decrease in chick hatching, although there were no developmental abnormalities or evidence of a dose response associated with the effect (Saito and others 1991; Saito and Suzuki 1995). In a series of studies in another laboratory, quail eggs were exposed to RF at 2450 MHz (Braithwaite and others 1991, Gildersleeve and others 1987; Galvin and others 1981; McRee and others 1983; Inouye and others 1982). There were no observed effects on hatching, body weight, malformations, or hematologic parameters at power density levels (~14 W/kg) that maintained a temperature of 37°C, although changes could be observed with exposures producing higher temperatures.

Development and Differentiation

Several recent studies have been undertaken to address the potential influence of RF exposure on developmental and differentiation processes. Lary and Conover (1987) completed a literature review of all teratogenesis studies extant to that date, including all studies with exposures from 300 to 3000 MHz, and below the ANSI exposure limit of 0.4 W/kg. They found no reports of teratogenic effects in the absence of organismal heating. However, Saito and Suzuki (1995), in a study involving exposure of chick embryos to a 428 MHz, 5.5 mW/cm^2 incident field, showed severely-delayed development according to the Hamburger-Hamilton staging. Ten replications of 10 eggs were completed, with developmental anomalies observed at SARs in the range of 8.6 mW/kg to 47.1 mW/kg.

Klug and others (1997) examined the effects of 150 MHz RF with exposures of CW, and modulated at 16, 60, and 120 Hz, on the development and differentiation of rat embryos. Using SARs of 0.2, 1.0, and 5.0 W/kg, no statistical differences between exposed and control embryos were observed following 48 hours of exposure. A few sporadic changes were observed under some exposure conditions, specifically increases in somite numbers at modulation frequencies of 16 and 120 Hz.

Reproduction

Temporary sterility was reported to occur in male rats exposed to 1.3 GHz at levels sufficient to cause intra-testicular temperatures of ~40°C (Lebovitz and others 1983, 1987). Longer-lasting alterations of reproductive efficiency have been observed with RF exposures producing temperatures greater than 45°C in animal testes. In several studies examining chickens exposed to moderately high levels of RF, the exposure was determined to cause a slight decrease in number of eggs laid (Krueger and others 1975; Giarola and Krueger 1974). However, in these studies, thermal parameters were not well characterized. A study at 900 MHz by Dasdag and others (1999) reported that exposure to RF for 1 month at 0.141 W/kg resulted in decreased diameter of seminiferous tubes, but no other effects were observed in a histological examination of all major organ tissues. There is some question with regard to the non-thermal nature of the observed effect since the study also reported an increase in rectal temperature. An additional report indicates an effect of low-level RF exposure on reproductive competance in rodents (Magras and Xenos 1997), although significant flaws in study design and control make the interpretation difficult.

Summary: Reproduction and Development

Although a number of studies have reported teratogenic effects of RF exposure in animal models, the positive results have almost always come from studies where significant temperature increases were observed in the dam. The few studies in which adverse effects on reproduction and/or development have been reported from non-thermal levels of RF exposure are relatively isolated. In addition, there appear to be possible species-specific differences in teratogenic responses to RF exposure suggesting that extrapolation of animal data to humans may not be straightforward. Examined in total, the available literature does not indicate any consistent effect of acute or chronic RF exposure on reproduction and development in animals unless significant temperature increases are produced. There are no human laboratory studies addressing this area of investigation.

IN VIVO STUDIES: CONCLUSIONS

Relatively few human laboratory studies investigating the RF exposure for possible health effects have been performed. Almost all of these were conducted at fairly low intensity levels using frequencies relevant to cellular telephony. Results have been mixed in the areas of RF effects on sleep and other behaviors and brain activity as determined by EEG. The range of effects, however, has been inconsistent from study to study without robust replication or direct implications for health impacts. Futhermore, studies examining RF effects on hormone or immune parameters have almost universally been negative.

With animal studies, across a wide range of biologic endpoints, a considerable number of studies have been conducted to evaluate the impact of RF exposure. The majority of laboratory studies have been performed at 2450 MHz, which is outside the frequency window of interest as defined for the purposes of this PAVE PAWS study (see Appendix A). However, in some cases, because little biological information was available from animals exposed at lower frequencies, some review of 2450 MHz data was utilized. The strong indication from the collected data suggests that various biological effects occur but generally only at field intensities where temperature is elevated in the biological target. Effects at non-thermal levels are seen only infrequently and seldom have they been independently replicated. Nonetheless, certain studies were sufficiently large, relevant to PAVE PAWS, and demonstrated robust and significant effects, suggesting that replication would be merited. Specifically, the Air Force-funded studies on blood-borne endpoints by Toler and others (1988) indicating depressed dopamine levels (and not to be confused with the Toler cancer studies noted earlier) deserve to be replicated utilizing exposure patterns more representative of the PAVE PAWS system than those used by Toler and others.

REFERENCES

Abhold, R.H., M.J. Ortner, M.J. Galvin, and D.I. McRee. 1981. Studies on acute in vivo exposure of rats to 2450-MHz microwave radiation. II. Effects on thyroid and adrenal axes hormones. Radiat Res 88:448-455.

Adey, W.R., C.V. Byus, C.D. Cain, R.J. Higgins, R.A. Jones, C.J. Kean, N. Kuster, A. MacMurray, R.B. Stagg, and W. Haggren. 1999. Spontaneous and nitrosourea-induced primary tumors of the central nervous system in Fischer 344 rats chronically exposed to 836 MHz modulated microwaves. Radiat Res 152:293-302.

Adey, W.R., C.V. Byus, C.D. Cain, R.J. Higgins, R.A. Jones, C.J. Kean, N. Kuster, A. MacMurray, R.B. Stagg, and G. Zimmerman. 2000. Spontaneous and nitrosourea-induced primary tumors of the central nervous system in Fischer 344 rats chronically exposed to frequency-modulated microwave fields. Cancer Res 60:1857-1863.

Akyel, Y., E.L. Hunt, C. Gambrill, and C. Vargas, Jr. 1991. Immediate post-exposure effects of high-peak power microwave pulses on operant behavior of Wistar rats. Bioelectromagnetics 12:183-195.

Albert, E.N., M.F. Sherif, N.J. Papadopoulos, F.J. Slaby, and J. Monahan. 1981a. Effects of nonionizing radiation on the purkinje cells of the rat cerebellum. Bioelectromagnetics 2:247-257.

Albert, E.N., M.F. Sherif, and N.J. Papadopoulos. 1981b. Effects of nonionizing radiation on the purkinje cells of the uvula in squirrel monkey cerebellum. Bioelectromagnetics 2:241-246.

Anane, R., M. Geffard, M. Taxik, D. Bodet, B. Billandel, P.E. Dulou, and B. Veyret. 2003. Effects of GSM-900 MHz microwaves on the experimental allergic encephalomyelitis (EAE) rat model. Bioelectromagnetics 24:211-213.

Anane R., P.E. Dulou, M. Taxile, M. Geffard, F. Crespeau, and B. Veyret. 2003. Effects of GSM-900 microwaves on female rats bearing DMBA-induced tumours. Radiat Res 160:492-497.

ANSI (American National Standards Institute). 1982. Safety Levels with Respect to Radio Frequency Electromagnetic Fields, 300 kHz to 100 GHz. Report No. ANSI C95.1 –1982. New York, NY: The Institute of Electrical and Electronic Engineers, Inc.

ANSI. 1992. IEEE Standard for Safety Levels with Respect to Human Exposure to Radio Frequency Electromagnetic Fields, 3 kHz to 300 GHz. Report No. C95-1-1992, New York, NY: The Institute of Electrical and Electronic Engineers, Inc.

Bartsch, H., C. Bartsch, E. Seebald, F. Deerberg, K. Dietz, L.Vollrath, and D. Meeke. 2002. Chronic exposure to a GSM-like signal (mobile phone) does not stimulate the development of DMBA-induced mammary tumors in rats: results of three consecutive studies. Radiat Res 157:183-190.

Bawin, S.M., R.J. Gavalas-Medici, and W.R. Adey. 1973. Effects of modulated very high frequency fields on specific brain rhythms in cats. Brain Res 58:365-384.

Bennet, G.D., V.K. Mohl, and R.H. Finnell. 1990. Embryonic and maternal heat shock responses to a teratogenic hyperthermic insult. Reprod Toxicol 4:113-119.

Berman, E., C. Weil, P. Phillips, H.B. Carter, and D. House. 1992. Fetal and maternal effects of continual exposure of rats to 970-MHz circularly-polarized microwaves. Electro Magnetobiol 11:43-54.

Bogolyubov, V.M., S.B. Pershin, I.D. Frenkel, V.D. Sidorov, A.I. Galenchik, Z.A. Sokolova, and I.B. Laprun. 1987. Immunobiological effect of bitemporal exposure of rabbits to microwaves. B Exp Biol Med 102:1118-1120.

Bogolyubov, V.M., S.M. Zubkova, I.D. Frenkel, Z.A. Sokolova, and I.B. Laprun. 1988. The functional state of thymus cells following microwave exposure of endocrine glands. Radiat Res 115:44-53.

Borbely, A.A., R. Huber, T. Graf, B. Fuchs, E. Gallmann, and P. Achermann. 1999. Pulsed high-frequency electromagnetic field affects human sleep and sleep eletroencephalogram. Neurosci Lett 275:207-210.

Boreham, D.R., J.A. Rolling, J. Misonoh, and R.E.J. Mitchel. 2002. Teratogenic effects of mild heat stress during mouse embryogenesis: effect of Trp53. Radiat Res 158:443-448.

Bornhausen, M. and H. Scheingraber. 2000. Prenatal exposure to 900 MHz, cell-phone electromagnetic fields had no effect on operant-behavior performances of adult rats. Bioelectromagnetics 21:566-574.

Bortkiewicz, A., B. Pilacik, E. Gadzicka, and W. Szymczak. 2002. The excretion of 6-hydroxymelatonin sulfate in healthy young men exposed to electromagnetic fields emitted by cellular phone—an experimental study. Neuroendocrinol Lett 23:88-91.

Braune, S., C. Wrocklage, J. Raczek, T. Gailus, and C.H. Lucking. 1998. Resting blood pressure increase during exposure to a radio-frequency electromagnetic field. Lancet 351:1857-1858.

Braune, S., A. Riedel, J. Schulte-Monting, and J. Raczek. 2002. Influence of a radiofrequency electromagnetic field on cardiovascular and hormonal parameters of the autonomic nervous system in healthy individuals. Radiat Res 158:352-356.

Braithwaite, L., W. Morrison, L. Otten, and D. Pei. 1991. Exposure of fertile chicken eggs to microwave radiation 2.45 GHz, CW during incubation: technique and evaluation. J. Microwave Power EE 26:206-214.

Brown, D.O., S.T. Lu, and E.C. Elson. 1994. Characteristics of microwave evoked body movements in mice. Bioelectromagnetics 15:143-161.

Cassel, J.C., B. Cosquer, R. Galani, and N. Kuster. 2004. Whole-body exposure to 2.45 GHz electromagnetic fields does not alter radial-maze performance in rats. Behav Brain Res 155:37-43.

Chagnaud, J.L., J.M. Moreau, and B. Veyret. 1999. No effect of short term exposure to GSM-modulated low-power microwaves on benzo(a)pyrene-induced tumours in rat. Int J Radiat Biol 75:1251-1256.

Chagnaud, J.L. and B. Veyret. 1999. 900 MHz (GSM) exposure on immune parameters in rats. Int J Radiation Biol 75:111-113.

Chang, B.K., A.T. Huang, W.T. Joines, and R.S. Kramer. 1982. The effect of microwave radiation 1.0 GHz on the blood-brain barrier in dogs. Radio Sci 17:165-168.

Chou, C.K., A.W. Guy, L.L. Kunz, R.B. Johnson, J.J. Crowley, and J.H. Krupp. 1992. Long-term low-level microwave irradiation of rats. Bioelectromagnetics 13:469-496.

Cleary, S.F., F. Nickless, L.M. Liu, and R. Hoffman. 1980. Studies of exposure of rabbits to electromagnetic pulsed fields. Bioelectromagnetics 1:345-352.

Cobb, B.L., J.R. Jauchem, and E.R. Adair. 2004. Radial arm maze performance of rats following repeated low level microwave radiation exposure. Bioelectromagnetics 25:49-57.

Cosquer B., R. Galani, N. Kuster, and J.C. Cassel. 2004. Whole-body exposure to 2.45 GHz electromagnetic fields does not alter anxiety responses in rats: a plus-maze study including test validation. Behav Brain Res: in press.

Croft, R.J., J.S. Chandler, A.P. Burgess, R.J. Barry, J.D. Williams, and A.R. Clarke. 2002. Acute mobile phone operation affects neural function in humans. Clin Neurophysiol 113:1623-1632.

D'Andrea, J.A., O.P. Gandhi, J.L. Lords, C.H. Durney, L. Astle, L.J. Stensaas, and A.A. Schoenberg. 1980. Physiological and behavioral effects of prolonged exposure to 915 MHz microwaves. J Microwave Power EE 15:123-135.

D'Andrea, J.A., E.R. Adair, and J.O. deLorge. 2003. Behavioral and cognitive effects of microwave exposure. Bioelectromagnetics Suppl 6:S39-S62.

Dasdag, S. 2000. Do cellular phones alter blood parameters and birth weight of rats. Electro Magnetobiol 19:107-113.

Dasdag, S. M.A. Ketani, Z. Akdag, A.R. Ersay, I. Sari, O.C. Demirtas, and M.S. Celik. 1999. Whole-body microwave exposure emitted by cellular phones and testicular function of rats. Urol Res 27:219-223.

Djordjevich, Z., N. Lazarevich, and V. Djokovich. 1977. Studies on the hematologic effects of long-term low-dose microwave exposure. Aviat Space Envir Med 1977:516-518.

Dubreuil, D., T.M. Jay, and B. Veyret. 2002. 900 MHz and 1800 MHz (GSM) exposure on spatial learning & memory in rats. Behav Brain Res 129:203-210.

Edelstyn, N. and A. Oldershaw. 2002. The acute effects of exposure to the electromagnetic field emitted by mobile phones on human attention. Neuroreport 13:119-121.

Eulitz, C., P. Ullsperger, G. Freude, and T. Elbert. 1998. Mobile phone modulate response patterns of human brain activity. Neuroreport 9:3229-3232.

Finnie, J.W., P.C. Blumbergs, J. Manavis, T.D. Utteridge, V. Gebski, and J.G. Swift. 2001. Effects of global system for mobile communication (GSM)-like radiofrequency fields on vascular permeability in mouse brain. Pathology 33:338-340.

Finnie, J.W., P.C. Blumbergs, J. Manavis, T.D. Utteridge, V. Gebski, and R.H. Davies. 2002. Effect of long-term mobile communication microwave exposure on vascular permeability in mouse brain. Pathology 34:344-347.

Frei, M.R., R.E. Berger, S.J. Dusch, V. Guel, J.R. Jauchem, J.H. Merritt, and M.A. Stedham. 1998. Chronic exposure of cancer-prone mice to low-level 2450 MHz radiofrequency radiation. Bioelectromagnetics 19:20-31.

Freude, G., P. Ullsperger, S. Eggert, and I. Ruppe. 1998. Effects of microwaves emitted by cellular phones on human slow brain potentials. Bioelectromagnetics 19:384-387.

Freude, G., P. Ullsperger, S. Eggert, and I. Ruppe. 2000. Microwaves emitted by cellular telephones affect human slow brain potentials. Eur J Appl Physiol 81:18-27.

Frey, A.H., S.R. Feld, and B. Frey. 1975. Neural function and behaviour: defining the relationship. Ann NY Acad Sci 247:433-439.

Frey, A.H. and J. Spector. 1986. Exposure to RF electromagnetic energy decreases aggressive behavior. Aggressive Behav 12:285-291.

Fritze, K., C. Sommer, B. Schmitz, G. Mies, K.A. Hossmann, M. Kiessling, and C. Wiessner. 1997. Effect of global system for mobile communication (GSM) microwave exposure on blood-brain barrier permeability in rat. Acta Neuropathol 94:465-470.

Galvin, M.J., D.I. McRee, C.A. Hall, J.P. Thaxton, and C.R. Parkhurst. 1981. Humoral and cell-mediated immune function in adult Japanese quail following exposure to 2.45 GHz microwave radiation during embryogeny. Bioelectromagnetics 2:269-278.

Giarola, A.J. and W.F. Krueger. 1974. Continuous exposure of chicks and rats to electromagnetic fields. IEEE T Microw Theory 22:432-437.

Gildersleeve, R.P., M.J. Galvin, D.I. McRee, J.P. Thaxton, and C.R. Parkhurst. 1987. Reproduction of Japanese quail after microwave irradiation 2.45 GHz CW during embryogeny. Bioelectromagnetics 8:9-21.

Gildersleeve, R.P., D.G. Satterless, D.I. McRee, T.E. Bryan, and C.R. Parkhurst. 1988. Plasma corticosterone in hemorrhaged Japanese quail after microwave irradiation in ovo. Comp Biochem Physiol 89:415-424.

Goldman, H., J.C. Lin, S. Murphy, and M.F. Lin. 1984. Cerebrovascular permeability to RB86 in the rat after exposure to pulsed microwaves. Bioelectromagnetics 5:323-330.

Gruenau, S., K.J. Oscar, M.T. Folker, and S.I. Rapoport. 1982. Absence of microwave effect on blood-brain barrier permeability to C14-sucrose in the conscious rat. Exp Neurol 75:299-307.

Guillet, R. and S.M. Michaelson. 1977. The effect of repeated microwave exposure on neonatal rats. Radio Sci 12:125-129.

Haarala, C., L. Bjornberg, M. Ek, M. Laine, A. Revonsuo, M. Koivisto, and H. Hamalainen. 2003. Effect of a 902 MHz electromagnetic field emitted by mobile phones on human cognitive function: a replication study. Bioelectromagnetics 24:283-288.

Heikkanen, P. and J. Juutilianen. 1999. Chronic exposure to 50-Hz magnetic fields or 900-MHz electromagnetic fields does not alter nocturnal 6-hydroxymelatonin sulfate secretion in CBA/S mice. Electro Magnetobiol 18:33-42.

Heikkinen, P., V.M. Kosma, T. Hongisto, H. Huuskonen, P. Hyysalo, H. Komulainen, T. Kumlin, T. Kumlin, T. Lahtinen, S. Lang, and J.Juutilainen. 2001. Effects of mobile phone radiation on x-ray induced tumorigenesis in mice. Radiat Res 156:775-785.

Hietanen, M., A.M. Hamalainen, and T. Husman. 2002. Hypersensitivity symptoms associated with exposure to cellular telephones: no causal link. Bioelectromagnetics 23:264-270.

Hietanen, M., T. Kovala, and A.M. Hamalainen. 2000. Human brain activity during exposure to radiofrequency fields emitted by cellular phones. Scan J Work Env Hea 26:87-92.

Higashiukubo, R., V.O. Culbreth, D.R. Spitz, M.C. LaRegina, W.F. Pickard, W.L. Straube, E.G. Moros, and J.L. Roti. 1999. Radiofrequency electromagnetic fields have no effect on the in vivo proliferation of the 9L brain tumor. Radiat Res 152:665-671.

Huber, R., T. Graf, K.A. Cote, L. Wittmann, E. Gallmann, D. Matter, J. Schuderer, N. Kuster, A.A. Borbely, and P. Achermann. 2000. Exposure to pulsed high-frequency electromagnetic field during waking affects human sleep EEG. Neuroreport 11:3321-3325.

Huber, R., V. Treyer, A.A. Borbely, J. Schuderer, J.M. Gottselig, H.P. Landolt, E. Werth, T. Berhold, N. Kuster, A. Buck, and P. Achermann. 2002. Electromagnetic fields, such as those from mobile phones, alter regional cerebral blood flow and sleep and waking EEG. J Sleep Res 11:289-295.

Huber, R., J. Schuderer, T. Graf, K. Jutz, A.A. Borbely, N. Kuster, and P. Achermann. 2003. Radio frequency electromagnetic field exposure in humans: estimation of SAR distribution in the brain, effects on sleep and heart rate. Bioelectromagnetics 24:262-276.

Imaida, K., M. Taki, T. Yamaguchi, T. Ito, S. Watanabe, K. Wake, A. Aimoto, Y. Kamimura, N. Ito, and T. Shirai. 1998. Lack of promoting effects of the electromagnetic near-field used for cellular phones (929.2 MHz) on rat liver carcinogenesis in a medium-term liver bioassay. Carcinogenesis 19:311-314.

Inaba, R., K Schishido, A. Okada, and T. Moroji. 1992. Effects of whole body microwave exposure on the rat brain contents of biogenic amines. Eur J Appl Physiol 65:124-128.

Inouye, M., N. Matsumoto, M.J. Galvin, and D.I. McRee. 1982. Lack of effect of 2.45 GHz microwave radiation on the development of preimplantation embryos of mice. Bioelectromagnetics 3:275-283.

Jensh, R.P. 1997. Behavioral teratologic studies using microwave radiation: is there an increased risk from exposure to cellular phones and microwave ovens? Reprod Toxicol 11:601-611.

Jensh, R.P., W.H. Vogel, and R.L. Brent. 1982a. Postnatal functional analysis of prenatal exposure of rats to 915 MHz microwave radiation. J Am Coll Toxicol 1:73-90.

Jensh, R.P., I. Weinberg, and R.L. Brent. 1982b. Teratologic studies of prenatal exposure of rats to 915 MHz microwave radiation. Radiat Res 92:160-171.

Kaplan, J., P. Polson, C. Rebert, K. Lunan, and M. Gage. 1982. Biological and behavioral effects of prenatal and postnatal exposure to 2450-MHz electromagnetic radiation in the squirrel monkey. Radio Sci 17:135-144.

Klug, S., M. Hetscher, S. Giles, S. Kohlsmann, and K. Kramer. 1997. The lack of effects of nonthermal RF electromagnetic fields on the development of rat embryos grown in culture. Life Sci 61:1789-1802.

Koivisto, M., C. Haarla, C.M. Krause, A. Revonsuo, M. Laine, and H. Hamalainen. 2001. GSM phone signal does not produce subjective symptoms. Bioelectromagnetics 22:212-215.

Koivisto, M., A. Revonsuo, C.M. Krause, C. Haarla, L. Sillanmaki, M. Laine, and H. Hamalainen. 2001. Effects of 902 MHz electromagnetic field emitted by cellular telephones on response times in humans. Neuroreport 11:413-415.

Krause, C.M., L. Sillanmaki, M. Koivisto, A. Haggqvist, C. Saarela, M. Laine, and H. Hamalainen. 2000a. Effects of electromagnetic field emitted by cellular phones on the EEG during a memory task. Neuroreport 11:761-764.

Krause, C.M., L. Sillanmaki, M. Koivisto, A. Haggqvist, C. Saarela, M. Laine, and H. Hamalainen. 2000b. Effects of electromagnetic fields emitted by cellular phones on the electroencephalogram during a visual working memory task. Int J Radiat Biol 76:1659-1667.

Krueger, W.F., A.J. Giarola, J.W. Bradley, and A. Shrekenhamer. 1975. Effects of electromagnetic fields on fecundity in the chicken. Ann NY Acad Sci 247:391-400.

Lai, H., A. Horita, and A.W. Guy. 1994. Microwave irradiation affects radial-arm maze performance in the rat. Bioelectromagnetics 15:95-104.

LaRegina, M., E.G. Moros, W.F. Pickard, W.L. Straube, J. Baty, and J.L. Roti. 2003. The effect of chronic exposure to 835.62 MHz FDMA or 847.74 MHz CDMA radiofrequency radiation on the incidence of spontaneous tumors in rats. Radiat Res 160:143-151.

Lary, J.M. and D.L. Conover. 1987. Teratogenic effects of radiofrequency radiation. IEEE Eng Med Biol Mag, March, 42.

Lary, J.M., D.L. Conover, E.D. Foley, and P.L. Hanser. 1982. Teratogenic effects of 27.12 MHz radiation in rats. Health Phys 29:381-383.

Lary, J.M., D.L. Conover, P.H. Johnson, and J.R. Burg. 1983. Teratogenicity of 27.12 MHz radiation in rats is related to duration of hyperthermic exposure. Bioelectromagnetics 4:249-255.

Lary, J.M., D.L. Conover, P.H. Johnson, and R.W. Hornung. 1986. Dose-response relationship between body temperature and birth defects in radiofrequency-irradiated rats. Bioelectromagnetics 7:141-149.

Lass, J., V. Tuulik, R. Ferenets, R. Rilsalo, and H. Hinrikus. 2002. Effects of 7 Hz-modulated 450 MHz electromagnetic radiation on human performance in visual memory tasks. Int J Radiat Biol 78:937-944.

Lebedeva, N.N., A.V. Sulimov, O.P. Sulimova, T.I. Korotkovskaya, and T. Gailus. 2001. Investigation of brain potentials in sleeping humans exposed to the electromagnetic field of mobile phones. Crit Rev Biomed Eng 29:125-133.

Lebovitz, R.M. 1981. Prolonged microwave irradiation of rats: effects on concurrent operant behavior. Bioelectromagnetics 2:169-186.

Lebovitz, R.M. and L. Johnson. 1983. Testicular function of rats following exposure to microwave radiation. Bioelectromagnetics 4:107-114.

Lebovitz, R.M., L. Johnson, and W.K. Samson. 1987. Effects of pulse-modulated microwave radiation and conventional heating on sperm production. J Appl Physiol 62:245-252.

Lebovitz, R.M. and R.L. Seaman. 1983. Pulse modulated and continuous wave microwave radiation yield equivalent changes in operant behavior of rodents. Physiol Behav 30:891-898.

Lee, T.M., S.M. Ho, L.Y. Tsang, S.H. Yang, L.S. Li, C.C. Chan, and S.Y. Yang. 2001. Effect on human attention of exposure to the electromagnetic field emitted by mobile phones. Neuroreport 12:729-731.

Liburdy, R.P. 1977. Effects of radio-frequency radiation on inflammation. Radio Sci 12:179-183.

Liburdy, R.P. 1979. Radiofrequency radiation alters the immune system: modulation of T- and B-lymphocyte levels and cell-mediated immunocompetence by hyperthermic radiation. Radiat Res 77:34-46.

Liburdy, R.P. 1980. Radiofrequency radiation alters the immune system. II. Modulation of in vivo lymphocyte circulation. Radiat Res 83:66-73.

Lin, J.C. and M.F. Lin. 1980. Studies on microwave and blood-brain barrier interaction. Bioelectromagnetics 1:313-323.

Lin, J.C. and M.F. Lin. 1982. Microwave hyperthermia-induced blood-brain barrier alterations. Radiat Res 89:77-87.

Lu, S.T., N.A. Lebda, S. Pettit, and S.M. Michaelson. 1985. The relationship of decreased serum thyrotropin and increased colonic temperature in rats exposed to microwaves. Radiat Res 104:365-386.

Lu, S.T., S.J. Lu, and S.M. Michaelson. 1986. Effects of microwaves on the adrenal cortex. Radiat Res 107:234-249.

Lu, S.T., N.A. Lebda, S. Pettit, and S.M. Michaelson. 1987. Effects of microwaves on three different strains of rats. Radiat Res 110:173-191.

Lu, S.T., D.O. Brown, C.E. Johnson, S.P. Mathur, and E.C. Elson. 1992. Abnormal cardiovascular responses induced by localized high power microwave exposure. IEEE T Bio-Med Eng 39:484-492.

Lyle, D.P., P. Schechter, W.R. Adey, and R.L. Lundak. 1983. Suppression of T-lymphocyte cytotoxicity following exposure to sinusoidally amplitude-modulated fields. Bioelectromagnetics 4:281-292.

Magras, I.N. and T.D. Xenos. 1997. RF radiation-induced changes in the prenatal development of mice. Bioelectromagnetics 18:455-461.

Mann, K. and J. Roschke. 1996. Effects of pulsed high-frequency electromagnetic fields on human sleep. Neuropsychobiology 33:41-47.

Mann, K. and J. Roschke. 2004. Sleep under exposure to high frequency electromagnetic fields. Sleep Med Rev 8:95-107.

Mann, K., P. Wagner, G. Brunn, F. Hassan, C. Hiemke, and J. Roschke. 1998. Effects of pulsed high-frequency electromagnetic fields on the neuroendocrine system. Neuroendocrinology 67:139-144.

Marcickiewicz, J., B. Chazan, T. Niemiec, G. Sokolska, M. Troszynski, M. Luczak, and S. Szmigielski. 1986. Microwave radiation enhances teratogenic effect of cytosine arabinoside in mice. Biol Neonate 50:75-82.

Mausset, A., R. deSeze, F. Mountpeyroux, and A. Privat. 2001. Effects of radiofrequency exposure on the GABAergic system in the rat cerebellum: clues from semi-quantitative immunochemistry. Brain Res 912:33-46.

McRee, D.I., J.P. Thaxton, and C.R. Parkhurst. 1983. Reproduction in male Japanese quail exposed to microwave radiation during embryogeny. Radiat Res 96:51-58.

Merritt, J.H., A.F. Chamness, and S.J. Allen. 1978. Studies on blood-brain barrier permeability after microwave-radiation. Radiat Environ Bioph 15:367-377.

Merritt, J.H., K.A. Hardy, and A.F. Chamness. 1984. In utero exposure to microwave radiation and rat brain development. Bioelectromagnetics 5:315-322.

Michaelson, S.M., R.A.E. Thomson, and J.W. Howland. 1961. Physiological aspects of microwave irradiation of mammals. Am J Physiol 201:351-356.

Mitchell, D.S., W.G. Switzer, and E.L. Bronaugh. 1977. Hyperactivity and disruption of operant behavior in rats after multiple exposures to microwave radiation. Radio Sci 12:263-271.

Mitchell, C.L., D.I. McRee, N.J. Peterson, H.A. Tilson, M.G. Shandala, M.I. Rudnev, V.V. Varetskii, and M.I. Navkatikyan. 1981. Results of a United States and Soviet Union joint project on nervous system effects of microwave radiation. Environ Health Persp 81:201-209.

Mitchell, C.L., D.I. McRee, N.J. Peterson, and H.A. Tilson. 1988. Some behavioral effects of short-term exposure of rats to 2.45 GHz microwave radiation. Bioelectromagnetics 9:259-268.

Moriyama, E., M. Salcman, and R.D. Broadwell. 1991. Blood-brain barrier alteration after microwave-induced hyperthermia is purely a thermal effect. I. Temperature and power measurements. Surg Neurol 35:177-182.

NCRP (National Council on Radiation Protection and Measurements). 1986. Biological Effects an Exposure Criteria for Radiofrequency Electromagnetic Fields, NCRP Report 86. Bethesda, MD: National Council on Radiation Protection.

Neilly, J.P. and J.C. Lin. 1986. Interaction of ethanol and microwaves on the blood-brain barrier of rats. Bioelectromagnetics 7:405-414.

Ohmoto, Y., H. Fujisawa, T. Ishikawa, H. Koizumi, T. Matsuda, and H. Ito. 1996. Sequential changes in cerebral blood flow, early neuropathological consequences and blood-brain barrier disruption following radiofrequency-induced localized hyperthermia. Int J Hyperther 12:321-334.

Oscar, K.J. and T.D. Hawkins. 1977. Microwave alteration of the blood-brain barrier system of rats. Brain Res 126:281-293.

Oscar K.J., S.P. Gruenau, M.T. Folker, and S.I. Rapoport. 1981. Local cerebral blood flow after microwave exposure. Brain Res 204:220-225.

Persson, B.R., L.G. Salford, and A. Brun. 1997. Blood-brain barrier permeability in rats exposed to electromagnetic fields used in wireless communication. Wirel Netw 3:455-461.

Persson, B.R., L.G. Salford, A. Brun, and L. Malmgren. 1992. Increased permeability of the blood-brain barrier induced by magnetic and electromagnetic fields. Ann NY Acad Sci 649:356-358.

Prausnitz, S. and C. Susskind. 1962. Effects of chronic microwave irradiation on mice. IRE T Biomed Electr 104-108.

Preece, A.W., G. Iwi, A. Davies-Smith, K. Wesnes, S. Butler, E. Lim, and A. Varey. 1999. Effect of a 915 MHz simulated mobile phone signal on cognitive function in man. Int J Radiat Biol 75:447-456.

Preston, E. and G. Prefontaine. 1980. Cerebrovascular permeability to sucrose in the rat exposed to 2,450-MHz microwaves. J Appl Physiol: Respiratory Environ Excercise Physiol 49:218-223.

Preston, E., E.J. Vavasour, and H.M. Assenheim. 1979. Permeability of the blood-brain barrier to mannitol in the rat following 2450 MHz microwave irradiation. Brain Res 174:109-117.

Radon, K., D. Parera, D.M. Rose, D. Jung, and L. Vollrath. 2001. No effects of pulsed radio frequency electromagnetic fields on melatonin, cortisol, and selected markers of the immune system in man. Bioelectromagnetics 22:280-287.

Repacholi, M.H., A. Basten, V. Gebski, D. Noonan, J. Finnie, and A.W. Harris. 1997. Lymphomas in Eμ–*Pim1* transgenic mice exposed to pulsed 900 MHz electromagnetic fields. Radiat Res 147:631-640.

Roberts, N.J. and S.M. Michaelson. 1983. Microwaves and neoplasia in mice: analysis of a reported risk. Health Phys 44:430-433.

Roschke, J. and K. Mann. 1997. No short-term effects of digital mobile radio telephone on the awake human electroencephalogram. Bioelectromagnetics 18:172-176.

Sagan, P.M. and R.G. Medici. 1979. Behavior of chicks exposed to low-power 450-MHz fields sinusoidally modulated at EEG frequencies. Radio Sci 14:239-245.

Saito, K. and K. Suzuki. 1995. Maldevelopment of early chick embryos induced by non-thermogenic dose radio frequency radiation at 428 MHz for the first 48 hours. Cong Anom 35:275-283.

Saito, K., K. Suzuki, and S. Motoyoshi. 1991. Lethal and teratogenic effects of long-term low-intensity radio frequency radiation at 428 MHz on developing chick embryo. Teratology 43:609-614.

Salford, L.G., A. Brun, J.L. Eberhardt, L. Malmgren, and B.R.R. Persson. 2003. Nerve cell damage in mammalian brain after exposure to microwaves from GSM mobile phones. Environ Health Persp 111:881-883.

Salford, L.G., A. Brun, J.L. Eberhardt, and B.R.R. Persson. 1993. Experimental studies of brain tumour development during exposure to continuous and pulsed 915 MHz radiofrequency radiation. Bioelectroch Bioener 30:313-318.

Salford, L.G., A. Brun, K. Sturesson, J.L. Eberhardt, and B.R.R. Persson. 1994. Permeability of the blood-brain barrier induced by 915 MHz electromagnetic radiation, continuous wave and modulated at 8, 16, 50, and 200 Hz. Microsc Res Techn 27:535-542.

Sandstrom, M., J. Wilen, G. Oftendal, and K.H. Mild. 2001. Mobile phone use and subjective symptoms. Comparison of symptoms experienced by users of analogue and digital mobile phones. Occup Med (Lond) 51:25-35.

Schmidt, R.E., J.H. Merritt, and K.H. Hardy. 1984. In utero exposure to low-level microwaves does not affect rat fetal development. Int J Radiat Biol 46:383-386.

Sienkiewicz, Z.J., R.P. Blackwell, R.G. Haylock, R.D. Saunders, and B.L. Cobb. 2000. Low-level exposure to pulsed 900 MHz microwave radiation does not cause deficits in the performance of a spatial learning task in mice. Bioelectromagnetics 21:151-158.

Smialowicz, R.J., J.S. Ali, E. Berman, S.J. Bursian, J.B. Kinn, C.G. Liddle, L.W. Reiter, and C.M. Weil. 1981a. Chronic exposure of rats to 100 MHz (CW) radiofrequency radiation: assessment of biological effects. Radiat Res 86:488-505.

Smialowicz, R.J., C.M. Weil, P. Marsh, M.M. Riddle, R.R. Rogers, and B.F. Rehnberg. 1981b. Biological effects of long-term exposure of rats to 970-MHz radiofrequency radiation. Bioelectromagnetics 2:279-284.

Smialowicz, R.J., M.M. Riddle, R.R. Rogers, and G.A. Stott. 1982b. Assessment of immune responsiveness of mice irradiated with continuous wave or pulse-modulated 425 MHz radio frequency radiation. Bioelectromagnetics 3:467-470.

Smialowicz, R.J., M.M. Riddle, C.M. Weil, P.L. Brugnolotti, and J.B. Kinn. 1982a. Exposure of rats to 425-MHz CW radiofrequency radiation: effects on lymphocytes. J Microwave Power EE 17:211-221.

Smythe, J.W. and B. Costall. 2003. Mobile phone use facilitates memory in male, but not female, subjects. Neuroreport 14:243-246.

Spalding, J.F., R.W. Freyman, and L.M. Holland. 1971. Effects of 800 MHz electromagnetic radiation on body weight, activity, hematopoiesis, and life span in mice. Health Phys 20:421-424.

Sutton, C.H., Q. Balzano, O. Garay, and F.B. Carroll. 1982. 452 MHz exposure on blood brain barrier permeability in pigs. J Microwave Power EE 17:280-281.

Sutton, C.H. and F.B. Carroll. 1979. Effects of microwave-induced hyperthermia on the blood-brain barrier of the rat. Radio Sci 14:329-334.

Sykes, P.J., B.D. McCallum, M.J. Bangay, A.M. Hooker, and A.A. Morley. 2001. Effect of radiofrequency exposure on intrachromosomal recombination in mutation and cancer. Radiat Res 156:495-502.

Takashima, S. and T. Asakura. 1983. Desickling of sickled erythrocytes by pulsed radio-frequency field. Science 220:411-413.

Takashima, S., B. Onaral, and H.P. Schwan. 1979. Effects of modulated RF energy on the EEG of mammalian brains: effects of acute and chronic irradiations. Radiat Environ Biophys 16:15-27.

Tahvanainen, K., J. Nino, P. Halonen, T. Kuusela, T. Laitinen, E. Lansimies, J. Hartikainen, M. Hietanen, and H. Lindholm. 2004. Acute heart rate and blood pressure responses during cellular phone use. Bioelectromagnetics 25:73-83.

Thuroczy, G., G. Kubinyi, M. Bodo, J. Bakos, and L.D. Bakos. 1994. Simultaneous response of brain electrical activity (EEG) and cerebral circulation (REG) to microwave exposure in rats. Rev Environ Health 10:135-148.

Tofani, S., G. Agnesod, P. Ossola, S. Ferrini, and R. Bussi. 1986. Effects of continuous low-level exposure to radiofrequency radiation on intrauterine development in rats. Health Phys 51:489-499.

Toler, J., V. Popovich, S. Bonasera, P. Popovich, C. Honeycutt, and D. Sgoutas. 1988. Long-term study of 435 MHz radiofrequency radiation on blood-borne end points in canulated rats—Part II: methods, results and summary. J Microwave Power EE 23:105-136.

Toler, J.C., W.W. Shelton, M.R. Frei, J.H. Merritt, M.L. Meltz, and M.A. Stedham. 1997. Long-term, low-level exposure of mice prone to mammary tumors to 435 MHz radiofrequency radiation. Radiat Res 148:227-234.

Utteridge, T.D., V. Gebski, J.W. Finnie, B. Vernon-Roberts, and T.R. Kuchel. 2002. Long-term exposure of Eυ–*Pim1* transgenic mice to 898.4 MHz microwaves does not increase lymphoma incidence. Radiat Res 158:357-364.

Vollrath, L., R. Spessert, T. Kratzsch, M. Keiner, and H. Hollmann. 1997. No short-term effects of high-frequency electromagnetic fields on the mammalian pineal gland. Bioelectromagnetics 18:376-387.

Vorobyov, V.V., A.A. Galchenko, N.I. Kukushkin, and I.G. Akoev. 1997. Effects of weak microwave fields amplitude modulated at ELF on EEG of symmetric brain areas in rats. Bioelectromagnetics 18:293-298.

Wagner, P., J. Roschke, K. Mann, W. Hiller, and C. Frank. 1998. Human sleep under the influence of pulsed radiofrequency electromagnetic fields: a polysomnographic study using standardized conditions. Bioielectromagnetics 19:199-202.

Wagner, P., J. Roschke, K. Mann, J. Fell, W. Hiller, C. Frank, and M. Grozinger. 2000. Human sleep EEG under the influence of pulsed radiofrequency electromagnetic fields. Results from polysomnographies using submaximal high power flux densities. Neuropsychobiology 42:207-212.

Ward, T.R. and J.S. Ali. 1985. Blood-brain barrier permeation in the rat during exposure to low-power 1.7 GHz microwave radiation. Bioelectromagnetics 6:131-143.

Ward, T.R., J.A. Elder, M.D. Long, and D. Svendsgaard. 1982. Measurement of blood-brain barrier permeation in rats during exposure to 2450-MHz microwaves. Bioelectromagnetics 3:371-383.

Williams, W.M., M. del Cerro, and S.M. Michaelson. 1984a. Effect of 2450 MHz microwave energy on the blood-brain barrier to hydrophilic molecules. B. Effect on the permeability to HRP. Brain Res 7:171-181.

Williams, W.M., W. Hoss, M. Formaniak, and S.M. Michaelson. 1984b. Effect of 2450 MHz microwave energy on the blood-brain barrier to hydrophilic molecules. A. Effect on the permeability to sodium fluorescein. Brain Res 7:165-170.

Williams, W.M., S.T. Lu, M. del Cerro, and S.M. Michaelson. 1984c. Effect of 2450 MHz microwave energy on the blood-brain barrier to hydrophilic molecules. D. Brain temperature and blood-brain barrier permeability to hydrophilic tracers. Brain Res 7:191-212.

Williams, W.M., J. Platner, and S.M. Michaelson. 1984d. Effect of 2450 MHz microwave energy on the blood-brain barrier to hydrophilic molecules. C. Effect on the permeability to [C-14] sucrose. Brain Res 7:183-190.

Yamaguchi, J., G. Tsurita, S. Ueno, S. Watanabe, K. Wake, M. Taki, and H. Nagawa. 2003. 1439 MHz pulsed TDMA fields affect performance of rats in a T-maze only when body temperature is elevated. Bioelectromagnetics 24:223-230.

Zook, B.C. and S.J. Simmens. 2001. The effects of 860 MHz radiofrequency radiation on the induction or promotion of brain tumors and other neoplasms in rats. Radiat. Res. 155:572-583.

Zwamborn, A.P., S.H.. Vossen, B.J. van Leersum, M.A. Ouwens, and W.N. Makel. 2003. Effects of global communication system radio-frequency fields on well being and cognitive functions of human subjects with and without subjective complaints. Netherlands Organization for Applied Scientific Research (TNO), FEL-03-C148.

8

Epidemiological Studies of the Possible Adverse Health Effects of Pulsed Radar Emissions

In this chapter we discuss a limited range of studies that addressed potential health effects among populations exposed to pulsed-radiofrequency wavelengths. Ideally, the best studies would be carried out in the fields of radar stations. Unfortunately, such studies are very limited in number and quality. One of the reasons is the difficulty of determining the precise dosage of the radar waves received by the population over a long period of exposure. As a result, epidemiological studies in such populations are few in number. When such studies are undertaken, they usually occur after long exposures and often within the context of occupational exposures. A large number of such studies deal with implied exposures due to assignment on death certificates such as "electrician."

We include in this chapter those studies which have been undertaken in situations where there have been serious concerns about possible health effects, such as in the case of the exposure in the Moscow Embassy (Lilienfeld and others 1978). In other instances there has been a defined cohort exposed to radar emissions of various types with follow-up for specific outcomes (Hill 1990; Robinette and others 1980; Groves and others 2002). We also include studies of Polish military personnel and of the region that was irradiated by the Skrunda radar facility, which report health effects of radar emissions not previously or subsequently confirmed by other observations.

EXPOSURES AT THE U.S. EMBASSY IN MOSCOW

A substantial RF exposure of employees working in the U.S. Embassy in Moscow (Lilienfeld and others 1978) occurred between 1953 and 1976. Those employees were exposed to microwave energy (500 MHz-3.0 GHz) at levels of 5 μW/cm^2 to 15 μW/cm^2 for periods of 9 to 18 hours per day. To determine the possible effects on morbidity and mortality, the health status of 1827 Department of State employees at the Moscow Embassy was compared with the health status of 2561 employees and their dependents at other U.S. Embassies in neighboring Eastern States using health records, health questionnaires, and death certificates. It was concluded that personnel working at the U.S. Embassy in Moscow from 1953 to 1976 suffered no ill effects from microwaves beamed at the building.

EXPOSURES AT THE MIT RADIATION LABORATORY

A cohort of 1592 white male physicists and engineers participating in a radar research and development project during Word War II was evaluated for mortality (Hill 1990). It was not possible to assign precise personal exposures although classes of exposure were assigned. Overall and cause-specific mortality, using standardized mortality ratios, was examined up until 1980 for an average survival of 36 years. Comparisons were made with all U.S. white males, and with a cohort of physicians. The MIT Radiation Laboratory experienced a lower than expected mortality based on a comparison with age-adjusted mortality in U.S. males, as well as with the comparison with the cohort of physicians. Among the different causes of death there was a slight elevation of rates for Hodgkin's disease and cirrhosis of the liver, neither of which reached statistical significance.

TWO STUDIES OF A U.S. NAVY COHORT

A total of 40,890 U.S. Naval personnel, believed to be exposed to radar energy while aboard navy ships, was studied for potential adverse health effects. No RF measurements were taken of the cohort, consisting only of men, who were categorized into high exposure and lower exposure based only upon job classification. This cohort was the subject of two separate investigations, including the original report by Robinette and others (1980) and subsequently by Groves and others (2002), which included an extended follow-up period.

In the initial investigation (Robinette and others 1980), the individuals were from U.S. Navy technical schools (1950-1954). Follow-up in this study was via hospital records, the Social Security Administration, and the National Death Index. Cause of death was obtained from death certificates. In the first study, there was a finding of significantly increased trauma mortality in the group with significantly elevated microwave exposure. On further examination, this increased mortality was the result of military aircraft accidents. Total mortality was not significantly different in the two groups. While mortality from cancer, stroke, chronic nephritis, influenza and pneumonia, and liver disease was elevated among the group with the higher level of exposure, none reached statistical significance.

The subsequent study of this cohort by Groves and others (2002), including a follow-up period of approximately 40 years, provided an opportunity to identify potential delayed health effects. In this study, comparisons of mortality were made between the two exposure groups and age-specific mortality rates in the U.S. white population. A total of 8393 deaths were identified for a cumulative crude mortality rate of 20.7%. The result for the comparison of the low radar exposure with that of the U.S. population was 0.80 (0.78-0.82) and for the high radar exposure rate it was 0.69 (0.67-0.71). This indicates that the whole cohort was in better health than the whole U.S. white male population of the same age distribution, an effect often referred to as the "healthy worker effect." The only causes of death with a statistically significant elevated SMR were for war-related deaths and for accidents involving air transportation in the high-exposure cohort (49.6 [27.5-89.6]) and (4.74 [3.89-5.76]), and war injuries were also significantly elevated in the low-exposure cohort (9.13 [2.28-36.5]). Non-significant elevations in the SMR were observed for lymphocytic leukemia (1.12 [0.69-1.83]) and non-lymphocytic leukemia (1.24 [0.90-1.69]) in the subgroup defined as high-potential exposure. There were some weaknesses in the studies consisting of lack of exposure measurements and potential losses in the follow-up which would have a tendency to lower SMRs. The authors of the more recent study conclude, "Radar exposures had little effect on mortality in this cohort of U.S. Navy veterans".

RADIO LOCATION STATION AT SKRUNDA

The Radio Location Station at Skrunda in Latvia was part of the Early Warning System of the Soviet Union (Romancuks 1996). The station had 2 pulsed-radar systems operating at 152 to 162 MHz, at a power of 1250 kW. The pulse duration was 0.8 msec with an interpulse interval of 41 msec and a pulse repetition rate of 24.5 Hz. The average-power level at nearby residential locations was less than 10 $\mu W/cm^2$ and the peak power level at these locations was 500 $\mu W/cm^2$. The radar systems were equipped with a pair of antennas measuring 250 × 12 m, which were angled slightly towards each other.

Selected children in the vicinity of the Skrunda Station underwent psychophysical testing. A total of 609 children aged 9-18 living within 20 km of the Skundra station were assumed to be exposed. The test battery (Polytest 8802), requiring 70 minutes to complete, including 11 tests that measure tapping rates, reaction to visual and auditory stimuli, attention, and memory (Kalmins and others 1996), was administered to 609 children. Among participants, 224 children lived where they were directly exposed, 385 children lived nearby but were not directly exposed, and an additional 357 children lived in non-exposed areas and served as controls. The authors concluded that children living in the exposed area performed worse on the tests than children living in the nearby area and even worse than the children in the control group. Many of the differences in performance were not statistically significant and many of the differences were found to show correlation differences of 0.25 and were barely significantly different from 0.0. In some categories performance on psychophysical tests appeared to be statistically impaired for children living in the radar line of sight compared to those living behind the radar and those living in a nearby community. The impairment occurred in all ages tested (9-18 years) in a tapping test, but in other cases the effect was only in one sex or in certain age groups. For instance, reaction times to auditory stimuli were significantly impared in ages 9-12 years, while ages 13-14 years showed no statistical difference on this test. The authors note that "evidence for a factor other than electromagnetic fields having caused the observed results was not found, but its existence cannot be ruled out . . .". The small populations exposed in the Skrunda radio stations do not make it likely that further epidemiological studies in this population can be mounted.

POLISH MILITARY STUDIES

Over the years 1971-1985 there was a study of cancer morbidity (first diagnosis of cancer) in a population consisting of career personnel in the armed forces of Poland (Szmigielski 1996). This population varied over the years from 118,500 to 142,200 with a mean of 127,800 (±9620). Based on military records, the authors designated a fraction of this population, which varied between 3400 and 3700 persons (mean 3720 ± 360), as exposed to RF/microwave radiation. This population was divided into 4 age-groups (20-29, 30-39, 40-49, and 50-59 years). The study consisted of a series of annual cross-sectional assessments of morbidity in a series of populations. This approach would be sensitive to changes in age makeup in each of the population groups over the period of study, or to possible changes in the exposure characteristics, but it also makes it difficult to compare the results of this study with conventional cohort studies that follow a particular cohort over a period of years. The exposed part of the population was exposed to pulse-modulated 150-3500 MHz radiation with 80-85% not exceeding 0.2 mW/cm^2 and 15-20% at 0.2-0.4 mW/cm^2. Individual exposures could not be determined. The total incidence of cancer diagnoses was determined by age groups for the total population and for the exposed population. Cancer occurrence was determined for the following sites: oral cavity, pharynx, esophagus and stomach, colorectal, liver/pancreas, lung, bones, skin including melanoma, kidney and prostate, brain and nervous system, thyroid, and hematopoietic and lymphatic systems. Cancer incidence in the non-exposed population was used as the expected rate for the total population. The overall cancer morbidity rate was 57.6 per 100,000 per year and the cancer morbidity rate was 119.1 per 100,000 per year in the exposed population. The gross observed/expected ratio therefore was 2.07 (1.12-3.58). The study reported significantly increased cancer morbidity in all exposed age groups except in 50-59-year-old subjects exposed to radiation. The largest subclass increases were from chronic myelocytic leukemia, acute myeloblastic leukemia, and non-Hodgkin's lymphoma (13.9, 8.62, and 5.82, respectively).

SUMMARY

The results of the studies cited here are mixed in characteristics. On the one hand are the results of studies of populations that are reasonably described in terms of radar exposure such as the population in the U.S. Embassy (Lilienfeld and others 1978). They are reinforced by similar exposures in the MIT radiation laboratory (Hill 1990), which also indicate a lack of health effect. In the MIT study the exposure characteristics were doubtful, but the duration of the exposure was early in 1942 until 1980. In all these studies there was uncertainty in the exposure classification. The total exposure in terms of people was small, but the follow-up was adequate to determine that there were negligible effects of exposure.

The next investigation was of a large cohort in the U.S. Navy (40,890 people) that was followed-up by two study groups that were relatively independent of each other (Robinette and others 1980; Groves and others 2002). In the first study, follow-up was from 1950 to 1974. The group was divided up into two cohorts, the first of which comprised 20,109 individuals who, based on their job description, had a higher exposure than the second cohort of 20,780 individuals, who received a much lower exposure. There were no direct exposure measurements made in either group after the initial assignment of dose. Neither investigation identified statistically significant associations with exposure and adverse health-related outcomes.

The committee has reviewed two studies that were connected with the Skrunda Radio Location System. This system was operating at a slightly lower frequency than PAVE PAWS but at a higher power level of 1250 kW. The first study reported a lower annual tree-ring width following the startup of the facility in 1971 and is reported in a previous chapter (Chapter 6 on biological effects). A second report addressed performance in a psychophysical test administered to youths of 9-18 years. The annual tree-ring width appeared affected (diminished) after the Skrunda Radio Location System became operational. In some catagories performance on psychophysical tests appeared to be statistically impaired for children living in the radar line of sight compared to those living behind the radar and those living in a nearby community (all ages tested 9-18 year-olds in a tapping test and ages 9-12 in reaction time to auditory stimuli; ages 13-14 showed no statistical difference on this latter test). Other tests were not statistically conclusive and the overall health impact on the children living in the line of sight to the beam, if any, is unclear.

The last study we report on is by Szmigielski (1996). Szmigielski studied cancer morbidity (first-cancer morbidity) in a varying population of 118,500 to 142,200 with a mean of 127,800 (127,800 $\pm$ 9620). The authors designated a fraction of this population as exposed to radar 150-3,500 MHz with 80-85% not exceeding 0.2 mW/cm^2 and 15-20% exposed to 0.2-0.6 mW/cm^2. The exposed population consisted of 3400 to 4600 men with a mean of 3720 $\pm$ 360. The cancer incidence in the non-exposed population was used as the background rate. For the whole population the SMR was 2.07 (1.12-3.58) for the observed/expected ratio, and the highest subclasses were for chronic myelocytic leukemia, acute myeloblastic leukemia, and non-Hodgkin's lymphoma (13.9, 8.62, and 5.82). The interpretation of this study and its relevance to PAVE PAWS is problematic because the radar exposure is $0.17 \times 3{,}720 \ (\pm\ 360) \times 0.4$ mW/cm^2 or 250 μV/cm^2, an amount considerably above the power densities experienced by the PAVE PAWS public. Because the 250 μV/cm^2 is elevated above the comparison levels and because the measurement of SMR is different from most of the other studies we have, the committee concludes that this finding is sufficiently different from the PAVE PAWS findings in Chapter 9 of this report that additional data would be required to determine its significance with respect to possible health effects resulting from exposures to PAVE PAWS.

REFERENCES

Groves, F.D., W.F. Page, G. Gridley, L. Lisimaque, P.A. Stewart, R.E. Tarone, M.H. Gail, J.D. Boice, and G.W. Beebe. 2002. Cancer in Korean War technicians: mortality survey after 40 years. Am J Epidemiol 155:810-818.

Hill, D. 1990. A longitudinal study of a cohort with past exposure to radar: the MIT radiation follow-up study (meeting abstract). Bioelectromagnetics Society, 12th Annual Meeting 10-14 June, San Antonio, TX Abstract No. A-2-6 p. 6.

Kalmins, T., R. Krizbergs, A.A. Kolodynski, and V.V. Kolodynska. 1996. Motor and psychological functions of school children living in the area of the Skrunda radio location station in Latvia. Sci Total Environ 180:87-93.

Lilienfeld, A.M., J. Tonascia, S. Tonascia, C.H. Libauer, G.M. Cauthen, J.A. Markowitz, and S. Weida. 1978. Foreign Service Study: Evaluation of Foreign Service and Other Employees from Selected Eastern European Posts. NTIS Document No. PB-28B 163/9GA p. 436.

Robinette, C.D., C. Silverman, and S. Jablon. 1980. Effects upon health of occupational exposure to microwave radiation (radar). Am J Epidemiol 112:39-53.

Romancuks, A. 1996. Measurement of the intensity of the electromagnetic radiation from the Skrunda Radio Location Station. Sci Total Environ 180 (1):51-56.

Szmigielski, S. 1996. Cancer morbidity in subjects occupationally exposed to high frequency (radio frequency and microwave) electromagnetic radiation. Sci Total Environ 180:9-17.

9

Health Considerations in the Cape Cod Population

INTRODUCTION

This chapter begins with a discussion of issues in epidemiologic research and then describes the status of disease on Cape Cod. This discussion is followed by an assessment of the sample size and statistical power needed to evaluate health risks of the PAVE PAWS radar. Finally the committee presents a geographical correlation study of health effects associated with exposure to the PAVE PAWS radar in the upper Cape Cod region surrounding the Massachusetts military reserve where PAVE PAWS is located.

The population characteristics of Cape Cod are important in interpretation of observed health effects and evaluation of epidemiological studies and are described briefly here.

Trend in Commuting

The trend in commuting to Boston has risen dramatically over the past 40 years (CCCR 2003). While the population of Barnstable county has tripled, off-Cape commuting has increased 1440 percent. In 2000, approximately 15% of Barnstable County's working residents commuted off-Cape to work, almost a 50% increase over the number commuting off-Cape in 1990. Nearly half the commuters went to the greater Boston Area.

Age Structure of the Population

The percent of the population on Cape Cod aged 65-plus rose from 22.8% in 1996 to 23.1% in 2000. This represents 36.7% of all households on Cape Cod.

Overall, 26% of Cape households have retirement income (US Census Bureau, Census 2000). Based on analysis of the 2000 Census, all 15 Cape towns saw growth in the 85-plus age group. Barnstable County led the 14 Massachusetts counties with the highest median age of 44.6. Furthermore, 11 Cape towns ranked in the top 20 oldest communities in Massachusetts.

The poverty level of Cape Cod is comparable to the state of Massachusetts as a whole. Overall, for families with children, the demographic poverty level is 3.4% for the state and 3.5% for Cape Cod. For those without children, the corresponding numbers are 2.2% and 2.1%.

Population Growth

Census data from 1980 and 1990 indicate that the growth of the state overall was 1.049% during this decade and for Barnstable County (Cape Cod) it was 1.26%. This reflects the trend of more retired elders moving to the Cape and a growing population commuting to jobs in the greater Boston area, in part with better intrastate road access.

The MISER data resource[1] used to project the annual and decade-level growth in future Massachusetts populations indicates that the actual recorded growth from 1990 to 1995 was 1.026% for the state of Massachusetts as a whole and 1.068% for Cape Cod. Furthermore, the projected growth from 1990 to 2000 was 1.053% for the decade for the state as a whole and 1.180% for Barnstable County.

Prevalence of Risk Factors on the Cape

The committee's access to historic data on risk factors was limited by their availability on BRFSS and insufficient numbers of responders to provide accurate estimates for counties. The Cape does have a lower prevalence of obesity than the state of Massachusetts as a whole through the 1990s (BRFSS data).

ISSUES IN EPIDEMIOLOGIC RESEARCH

Epidemiologic research can encompass a spectrum of strategies and methodological approaches to describe the occurrence of and risk factors for disease. Research designs can include descriptive, ecologic, case series, cross-sectional, case-control, cohort, and clinical trials. The type of information derived from these different strategies typically differs substantially with regard to the metrics obtained and the level of causal inference. Within the context of considering po-

[1]The Massachusetts Institute for Social and Economic Research (MISER) was founded in 1981 by the University of Massachusetts. MISER is an interdisciplinary research institute of the College of Social and Behavioral Sciences.

tential health effects from the PAVE PAWS facility, a combination of epidemiologic approaches can be recommended, which will provide information relevant to this issue.

As a general rule, most epidemiologic investigations are hypothesis-related. While the hypothesis focus can be hypothesis generating, the nature of epidemiologic research is typically hypothesis testing. Hypothesis generation is restricted to situations where there exists a lack of substantive data on an underlying mechanism of disease causation. Any observed associations are then evaluated for any existing or speculative mechanisms that may support the observation. One common outcome is the proposal of laboratory- or animal-based experiments that might provide confirmatory data relating to the observed association. Within a hypothesis-generation study, it is possible to define either the exposure(s) or the effect(s) of interest. For example, within a cohort-study design, multiple effects and/or exposure can often be addressed, depending on the cohort characterization and follow-up. In the situation of a case-control study, the hypothesis-generation component would be focused on exposure-specific factors. A subject of importance in research designated as hypothesis generating is the issue of multiple comparisons. That is, when researchers investigate many different factors and/or effects the probability of observing some significant associations increases due to random chance.

Hypothesis testing is within the context of available data to support a tenable biologic mechanism between a given exposure (e.g., PAVE PAWS-related) and a specific health effect. By definition, hypothesis-testing epidemiologic research specifies the exposure and outcome of interest as well as the direction of effect (i.e., increase of risk or protective effect).

Given the lack of sufficient supportive data to propose specific hypotheses to be tested (that is, supported by a reasonable and justified underlying biological rationale) the spectrum of health-related effects that could be considered within the context of the PAVE PAWS is almost unlimited. Thus, associations observed as the result of multiple comparisons must be viewed with extreme caution and considered tentative until reproduced in another population or supported by biologically based evidence. Some examples of possible health effects that might be considered within a hypothesis-generating exercise include cancer, cardiovascular disease, immunodeficiency, autoimmune disease, infertility, pregnancy outcomes, birth defects, attention deficit hyperactivity disorder, autism, depression, dementia, asthma, or endocrinopathies. The ability to conduct scientifically valid research is dependent on (1) the feasibility of complete ascertainment and characterization of disease state(s) of interest, many of which include a heterogeneous group of diseases, (2) a sufficient level of statistical power to detect levels of risk considered to be of importance, and (3) the ability to accurately characterize, in detail, the exposure of interest among the relevant population. Deficiencies in one or more of these attributes can invalidate the results and corresponding conclusions of epidemiologic research.

HEALTH ON CAPE COD

The health of the population can be monitored using a range of measures including the incidence or onset of new cases of disease, the prevalence or frequency of the condition in the population, and mortality, usually reported as the rate of death in the population. Other measures such as a person-years of life lost may be used to rank the impact of conditions on the population. The Massachusetts Department of Public Health produces reports that summarize both the burden of disease and the prevalence of risk factors for chronic conditions for the state as a whole. Data on cancer are reported using the resources of the state tumor registry and birth defects are reported using specialized data collection procedures. Many other conditions, however, are not routinely summarized either for the state as a whole or for smaller subdivisions such as counties or towns. Therefore we have available extensive data on cancer, more recent data on birth defects, and no comparable data on heart disease, stroke, or other major leading causes of disability and death.

Premature Mortality

The overall burden of illness for the state has been summarized in the measure of premature mortality (Eyles and Birch 1993). This is defined as the deaths occurring before age 75, and may be considered to reflect a general measure of the overall health of the population. While premature mortality reflects both the community-level health status and other correlates, such as access to care, it is important to note that the figures also include deaths from external causes (e.g., accidents, suicides, homicides, etc.). Moreover, premature mortality does not give us a source to identify why an area may be high or low, and because it is a general measure of early death, it may not precisely measure overall health. In fact, from time to time, it may obscure important disease burdens in subgroups of the population. The most recently reported data from Massachusetts draw on the reported deaths in 2001.[2] The state average premature mortality rate was 347 (95% confidence interval 342 to 352) deaths per 100,000 persons ages 0-74, after adjustment to the age distribution of the 2000 census. The premature-mortality rates for towns on upper Cape Cod are summarized in Table 9-1.

Of note, only one of the towns, Bourne, has a premature mortality rate in 2001 that is higher than the state average of 347 (342-352), though this difference is not statistically significant. Importantly, many factors can contribute to this measure. The reports summarized below focus on the incidence of cancer and birth defects, which are more proximate in time to any environmental exposure than is mortality, except in the situation of natural disasters associated with trauma and death.

[2]Premature Mortality 2001: http://www.state.ma.us/dph/bhsre/death/2001/premtrmortal2001.pdf.

TABLE 9-1 Age-Adjusted Premature Mortality Rates (deaths per 100,000) for Upper Cape Cod Towns

Town	Premature Mortality Rate (95% Confidence Interval)
Barnstable	275 (233-324)
Bourne	388 (307-489)
Falmouth	333 (276-400)
Mashpee	332 (241-454)
Sandwich	288 (218-379)

SOURCE: Premature mortality rates, Massachusetts Department of Public Health. (Premature Mortality 2001, www.state.ma.us/dph/bhsre/death/2001/pretrmortal2001.pdf).

Cancer

From the inception of the Massachusetts Cancer Registry and the initial analysis of cancer incidence by towns, there has been concern that the rates of cancer are elevated in the Cape Cod population. This has led to a range of analyses that the committee has summarized in this section. In December 1987, the Massachusetts Department of Public Health issued a request for proposals to respond to three important public health issues affecting the upper Cape Cod (Barnstable, Bourne, Falmouth, Mashpee, and Sandwich).

Apparent elevations in cancer incidence and mortality were observed for certain kinds of cancer as compared to the statewide averages and the lower Cape Cod region. Consistently elevated mortality rates for lung cancer and leukemia were seen for the towns of Falmouth and Bourne. In addition, since the inception of the Massachusetts Cancer Registry in 1982, statistically significant excesses were seen in the incidence of cancers of the breast, colon and rectum, lung, and blood forming organs, and statistically variable increases were seen for cancers of the pancreas, kidney, and bladder in at least one of the upper Cape towns.

Secondly, there were many known or suspected environmental hazards affecting the upper Cape. These included both groundwater and air contaminants from a variety of sources including the Massachusetts Military Reservation.

Third, there was substantial concern among citizen advocates who pressed forcefully and consistently for an in-depth evaluation of the relationship between the environmental hazards and cancer rates. The upper Cape Cod cancer-incidence study was conducted by Drs. Aschengrau and Ozonoff of Boston University School of Public Health and submitted to the department of public health in September 1991. These investigators undertook a population-based case-control study that evaluated the relationship between exposures to known or suspected environmental hazards and nine types of cancer. Cases consisted of newly diagnosed cancers in the years 1983 to 1986 among permanent upper Cape residents. The main environmental exposures considered were air and water pollution associated with the Massachusetts Military Reservation, Canal Electric Plant, the Barnstable airport, and other sources, perchlorethylene in water distribution system pipes, radiofrequency radiation from PAVE PAWS, electromagnetic radiation from 115 kV transmission lines and substations, and possible exposure to pesticides among residents who lived near cranberry bogs.

The study consisted of incident cancer of lung, n = 251, breast, n = 265, colon and rectum, n = 315, bladder, n = 62, kidney, n = 35, pancreas, n = 37, leukemia, n = 36, brain, n = 37, and liver, n = 4, as reported by the Massachusetts Cancer Registry in the years 1983 to 1986. Since many individuals were deceased by the start of the study, the living and deceased upper Cape residents were selected as controls (n = 1285). Trained interviewers queried all subjects or the next-of-kin either by telephone or in person to obtain demographic, occupational, and residential history and information on potential confounding variables such as cigarette smoking. Overall, approximately 81% of cases and 79% of controls were interviewed. The majority of the environmental exposure data was collected independently of the interview and linked to the study subjects using the residential histories.

Each exposure was examined separately in relation to all cancers combined, and then for the individual cancer sites. Most exposures were categorized as dichotomous variables, and then further analyzed according to distance, cumulative duration, and direction. The strength of the association between a particular exposure and each cancer site was measured by the odds ratio. Most analyses were conducted with and without latency to account for the possibility that the exposures under study could be either cancer initiators or a combination of both initiators and promoters.

Few study subjects had potential exposure to drinking-water contaminants from groundwater plumes emanating from various sites on or off the Massachusetts Military Reservation. Those plumes, including the Ashmet Valley plume, do not appear to account for much, if any, of the cancer burden to the population.

Results suggest approximately twofold elevations in the risk of brain cancer among those who ever had a residence supplied with public water on the upper Cape, particularly from the Barnstable Water Company. These findings cannot be considered conclusive because of limitations in the data. These limitations include the lack of details regarding the historical pattern of water contamination in the area and patterns of water use among the subjects, the large number of subjects dropped from these analyses because of missing data, and the almost complete overlap between exposures to the Barnstable Water Company and Barnstable airport for which an elevation in brain-cancer risk is also seen.

The investigators examined another public drinking-water exposure, perchlorethylene (PCE) from the water-distribution pipes, in relation to leukemia, bladder, and kidney cancer. A twofold increase in the risk of leukemia and bladder cancer among those supplied with water from pipes that leached PCE was observed. These risk estimates were not statistically stable, a reflection of small numbers. For these reasons, the results are consistent with the hazards of PCE contamination in some of the distribution systems of the upper Cape.

Among residents located in proximity to the gun and mortar position on the Massachusetts Military Reservation, the investigators observed associations between possible airborne exposure and the risk of lung and breast cancer. Subjects who lived closer to the gun and mortar positions had a modest increase in the risk of lung cancer. Likewise, there was an increase in risk of breast cancer among subjects who lived closer to the sites. Among residents exposed for more than 20 years, the investigators found an increase in the risk of lung and breast cancer. Non-meaningful elevations were seen for the other cancer sites.

While no association was seen for PAVE PAWS, the exposure data were inadequate for these investigators to draw sound conclusions (greatly enhanced power-density data and modeled power-density estimates became available in 2004 (BSL 2004)). Possible pesticide exposure associated with living near the cranberry bogs was also examined. Again, an increase in risk of brain cancer was seen among individuals who ever lived within half a mile of a bog. The risk remained elevated when subjects with other relevant exposures were excluded. Other potential environmental exposures were also considered, including electromagnetic radiation from transmission lines. Results suggest that extremely low-frequency electromagnetic fields may relate to increases in the risk of bladder cancer and breast cancer. An association was also observed for brain cancer among those who ever swam in Johns Pond and leukemia among those ever swimming in local ponds.

A comparison of the calculated SIR for 5 categories consisting of total cancers, breast, colon, lung, and prostate cancer during 1987-1994 versus 1995-1999 for 5 towns in upper Cape Cod (see Table 9-2) demonstrated no consistent pattern of increase during the latter time period. During these two time periods, a decrease was observed in 15 out of 25 SIRS, no change in 4 out of 25 SIRS, and an increase in 6 out of 25 SIRs, thus suggesting that increasing duration of the presence of the PAVE PAWS radar has not resulted in increased incidence of cancer. Analyses covering longer durations may be required to confirm this initial observation.

In summary, these results across the spectrum of potential environmental exposures point to numerous potential associations that require further investigation as the results were largely inconclusive. The limited number of cases included in the study for each cancer site preclude definitive conclusions.

It should be noted that multiple comparisons are an issue whenever a state tumor registry breaks out incidence data for every town. Data are presented for 351 cities and towns in the Commonwealth of Massachusetts. Rothman and Greenland in their text point to the limitations of using tumor registry data for such descriptive statistics. Smaller towns will have higher variability and be at

TABLE 9-2 Comparison of Standardized Incidence Ratios for Upper Cape Cod During Periods 1987-1994 vs. 1995-1999

	Total Cancer		Breast		Colon		Lung		Prostate	
Town	1987-1994	1995-1999	1987-1994	1995-1999	1987-1994	1995-1999	1987-1994	1995-1999	1987-1994	1995-1999
Barnstable	113	102	115	101	119	79	123	96	124	125
Bourne	114	114	117	101	110	118	123	109	116	114
Falmouth	122	114	119	118	120	101	116	106	154	121
Mashpee	121	151	101	142	113	137	109	155	178	196
Sandwich	109	97	119	110	99	99	104	68	134	106

SOURCE: Massachusetts Department of Public Health.

increased risk of reporting significant variation from the state average. Also, adjacent towns are at risk of increased (or decreased) cancer incidence because of autocorrelation among risk factors that are not accounted for in the age-standardized data. (Rothman and Greenland,1998).

The results from the upper Cape cancer-incidence study stimulated additional investigations, of both the health effects associated with environmental exposures from the Massachusetts Military Reservation, and the Silent Spring investigation of breast cancer. In addition, the Bureau of Environmental Health Assessment of childhood cancer incidence on Cape Cod commissioned a study from 1982 for 1994. These are reviewed in sequence.

MASSACHUSETTS MILITARY RESERVATION

Ryan and colleagues (MDPH 2002) employed extensive analytic techniques to explore the possible associations between the Massachusetts Military Reservation and health among upper Cape Cod residents. In particular, they observed that the female lung-cancer rates were higher than expected in the immediate upper Cape area, especially in the vicinity to the south, east, and west of the Massachusetts Military Reservation. Rates of lung cancer begin declining in regions further away from the Massachusetts Military Reservation, although there are isolated areas of apparently high rates. While lung-cancer rates were high in the upper Cape, particularly for females, the excesses occur over a broad area of approximately 20 to 30 square miles, making it unlikely that a single nearby point source of exposure is responsible.

With respect to birth weights, Ryan and colleagues extended analysis over a larger spatial area, and across a 7-year time frame, showed good consistency among different methodologic approaches. Unlike the cancer studies, the analysis showed that additional data added in this analysis attenuated the previous reported associations. Birth weights in the upper Cape appear now, on average, modestly higher than the statewide average. While there has been a downward trend in birth weights across the state of Massachusetts, some of this trend is explained in terms of changes in time periods, particularly an increase in the number of multiple births. The lower than expected birth weights found in the Wareham area and in a few other locations, including Mashpee in the later years, do not suggest any particular relationship with the Massachusetts Military Reservation.

The investigators concluded that lack of a consistent finding between the lung-cancer rates and birth-weight maps leads to the conclusion that although some exposures from the Massachusetts Military Reservation may have contributed to the induction of lung cancers this does not seem to lend support for an association with birth weights.

Breast Cancer

The Massachusetts Department of Public Health awarded funds to the Silent Spring Institute, which conducted a study of breast cancer in relation to environmental risk factors drawing on data from 1982 to 1990. The investigators noted that 9 of the 15 towns in Massachusetts with significantly elevated breast cancer were located on Cape Cod, for the period from 1982 to 1990. Furthermore, towns in the mid-Cape and lower Cape were among those with the highest incidence compared to the rest of the state. Five of these mid- and lower-Cape towns had incidence rates that were 25 to 55% higher than the state average. The Silent Spring study looked in substantial detail at the potential environmental contamination that may relate to risk of breast cancer, focusing on environmental estrogens, PCE exposure, DDT exposure, and the role of the Massachusetts Military Reservation as potential sources of exposure. The investigation concluded that breast-cancer rates on Cape Cod do not appear to be fully explained by the individual characteristics of the women (including, for example, age, family history, reproductive history, alcohol consumption, smoking, and age at onset of menstruation and menopause). A number of specific environmental facilities of concern including the PAVE PAWS radar system, the Massachusetts Military Reservation, Canal Electric Plant, and Pilgrim Nuclear Power Station were considered and fail to explain why breast-cancer rates are elevated Cape-wide. However, the investigators could not rule out localized effects from these various facilities.

Breast cancer is predominantly driven by female hormones (Colditz 1998; Willett and others 2000). While the studies of breast cancer on Cape Cod have considered a number of reproductive risk factors (Aschengrau and others 1998; Coogan and Aschengrau 1998; Coogan and Aschengrau 1999; Aschengrau and others 2003; McKelvey and othes 2004), it appears that in these analyses, the use of postmenopausal estrogens, which significantly increase the risk of postmenopausal breast cancer, were not included. The investigators did, however, note that the use of mammography could not explain the increase in incidence that was observed.

A detailed analysis of residence on Cape Cod and risk of breast cancer (McKelvey and others 2004) included some 1121 cases of breast cancer occurring between 1988 and 1995 and 992 controls. These cases and controls were drawn from the entire Cape. After adjusting for age, family history, parity, age at first birth, education, body mass index, and history of breast cancer, longer duration of residence on Cape Cod was associated with increased risk of breast cancer. In general, after excluding women with a history of breast cancer, those women who had lived 20 or more years on Cape Cod had significantly elevated risk. For example, those living 40 to 47 years on Cape Cod had an adjusted relative risk of 1.57 (95% CI 1.07 to 2.29). Although uncontrolled confounding cannot be ruled out, it is unlikely that duration of residence is related to use of postmenopausal hormones or other established risk factors. Importantly, this elevated risk of breast cancer is not limited to the upper Cape, or other defined sub-region on the Cape.

Upper Cape Cod Cancer-Incidence Review, 1986 to 1994

The Massachusetts Department of Public Health, Bureau of Environmental Health Assessment, reported the cancer rates for towns in Cape Cod through the "Upper Cape Cod Cancer Incidence Review, 1986-1994" released in June 1999. The study population for the statistical review included individuals who were residing in the Massachusetts towns of Barnstable, Bourne, Falmouth, Mashpee, and Sandwich between January 1986 and December 1994. Those five towns are collectively referred to as upper Cape Cod. The 1990 United States census data were used to represent the population of those towns for the year 1990. Cases were identified through the Massachusetts Cancer Registry. A total of 6244 primary-cancer cases were diagnosed between 1986 and 1994 among residents of the five upper Cape towns.

Cancer cases were selected for inclusion based on the reported address of residence, not the address of the diagnosing hospitals. Massachusetts residents diagnosed at out-of-state hospitals were also included in the series because of the Massachusetts Cancer Registry's reciprocal reporting agreements with states where Massachusetts residents may be diagnosed including Connecticut, Florida, Maine, New Hampshire, New York, and Rhode Island. It is estimated that the Massachusetts Cancer Registry files contain data on 90 to 95% of reportable cases.

To estimate the standardized incidence ratios (SIRs) for each town, the incidence in thc town is compared to the incidence for a given cancer in the state as a whole, which is used to represent the cancer experience of a comparable "normal" population. Thus the expected number of cases was estimated using standard approaches, within age-specific strata, of age, and then summed across age groups, to provide the overall expected number of cases for each town.

The descriptive analysis of cancer-incidence data for the upper Cape towns addressed 23 specific cancer types as well as total cancer incidence for the upper Cape as a whole, for each of the five towns, and for each of the 30 census tracts that compose the upper Cape. Cancer incidence for cancer types combined for the upper Cape as a whole was elevated significantly compared to the Massachusetts statewide experience. Cancer, however, is not one disease but rather a collection of different diseases, with distinct etiologies. The majority of cancers among upper-Cape residents were diagnosed in the lung, breast, prostate, and colorectal organs. These four cancer sites comprise slightly more than half of all cancers observed both on Cape Cod and in the United States overall. For the five upper-Cape towns combined, statistically significant elevations were observed for female breast cancer, colorectal cancer among males and both sexes combined, female lung cancer, and prostate cancer. In addition, statistically significant elevations were seen for melanoma. No statistically significant elevations for the upper Cape as a whole were seen in the other leading specific cancer types evaluated in this study (see Table 9-3).

The overall excess incidence of cancer in the upper Cape towns was 11% (SIR = 111; 95% CI 108-113). This was consistent for men (SIR = 112) and for women (SIR = 110).

It is of note that there is no common risk factor for breast, prostate, lung, and colorectal cancer other than increasing age. Breast and prostate cancers are largely hormonal in origin, cigarette smoking predominantly causes lung cancer, and colorectal cancer has many lifestyle factors that reflect a western lifestyle (including lack of physical activity, obesity, alcohol, low folate intake, height, etc.).

TABLE 9-3 Cancer Cases Diagnosed on Upper Cape Cod, 1986-1994, and Standardized Incidence Rate Ratios (SIR)

Type	Male	Female	Total	SIR	95% CI
Bladder	167	57	224	102	(89-116)
Brain	50	46	96	99	(80-121)
Breast	5	935	940	110	(103-117)
Cervical	0	62	62	118	(90-151)
Colon/rectum	491	427	918	112	(105-120)
Esophagus	42	21	63	97	(74-124)
Hodgkins	28	15	43	116	(84-156)
Kidney	81	57	138	106	(89-125)
Larynx	66	9	75	103	(81-129)
Leukemia	68	33	101	105	(85-127)
Liver	21	11	32	99	(68-139)
Lung & bronchus	497	417	916	112	(105-120)
Melanoma (skin)	96	76	172	135	(116-157)
Mouth and pharynx	96	52	149	101	(86-119)
Multiple myeloma	20	25	45	90	(66-121)
NHL	86	94	180	95	(82-110)
Ovary	0	101	101	100	(81-122)
Pancreas	62	66	128	112	(93-133)
Prostate	1001	0	1001	130	(122-139)
Stomach	67	38	105	93	(76-112)
Testis	33	0	33	122	(84-172)
Thyroid	12	35	47	107	(79-143)
Uterus	0	146	146	94	(79-110)
Other	221	201	422	102	(93-112)
Total	3210	2927	6137	111	(108-113)

SOURCE: Upper Cape Cod Cancer Incidence Review 1986-1994. Final Report Volume 2. Appendix C. Mass. Department of Public Health, Bureau of Environmental Health Assessment, Environmental Epidemiology Unit, June 1999.

TABLE 9-4 Standardized Incidence Ratios, 1995-1999, for Upper Cape Cod Towns

Town	Total Cancer	Breast (f)	Colon (total)	Lung (total)	Prostate (m)
Barnstable	102	101	79 ($p < 0.01$)	96	125 ($p < 0.001$)
Bourne	114 ($p < 0.01$)	101	118	109	114
Falmouth	114 ($p < 0.001$)	118 ($p < 0.05$)	101	106	121 ($p < 0.001$)
Mashpee	151 ($p < 0.001$)	142 ($p < 0.05$)	137	155 ($p < 0.001$)	196 ($p < 0.001$)
Sandwich	97	110	99	68 ($p < 0.001$)	106

SOURCE: Cancer Incidence In Massachusetts 1995-1999: City and Town Supplement.

This raises concern that the estimation of the expected number of cases of cancer may be distorted due to the population characteristics of Cape Cod residents, the distribution of age within any five-year age group, and the general construct of the definition of residents. For this definition of residence, for example, is a resident defined for the Massachusetts Cancer Registry the same as a resident defined for the United States Census?

The Massachusetts Department of Public Health updated information of cancer incidence and mortality with its report "Cancer Incidence and Mortality in Massachusetts 1996-2000." Through 2000, total cancer-mortality rates declined significantly while incidence rates varied according to cancer site. The City and Town Supplement reports the cancer data for towns and is most recently available for the period 1995-1999.[3] Data for upper Cape Cod towns are summarized in Table 9-4.

These updated data indicate that the incidence of cancer remains elevated in three of the five upper Cape towns, breast cancer in two, and prostate cancer in three. On the other hand, significantly lower than expected rates of colon cancer are observed in Barnstable, while lung cancer is significantly elevated in Mashpee but significantly reduced in Sandwich. While the small number of cases in each town may contribute to this variability, it is clear that the picture of elevated incidence across these four leading cancers has changed considerably in these more recent data.

Childhood Cancer

The Bureau of Environmental Health Assessment of the Massachusetts Department of Public Health released a report in September 1999 "Assessment of

[3]http://www.state.ma.us/dph/bhsre/mcr/99/supplement/tables.htm.

Childhood Cancer Incidence on Cape Cod, Massachusetts, 1982 to 1994." This report addressed cancer among children up to and including age 19 at the time of diagnosis on Cape Cod. All cancers diagnosed in children between 1982 and 1994 and reported through the Massachusetts Cancer Registry were included. The purpose of the evaluation was to discuss the types of cancer that were occurring among Cape Cod children and to identify areas with a greater than expected occurrence of childhood cancer. Like other health evaluations, this investigation was initiated at the request of Cape Cod citizens who were concerned that childhood cancer might be a problem on Cape Cod.

Each cancer case was geocoded to the location of the child's residence at the time of diagnosis. Standardized incidence ratios were calculated to estimate the number and types of cancers diagnosed among children, and determine if this was different from what was expected based upon statewide childhood cancer rates.

A total of 101 cancers were diagnosed among children on Cape Cod between 1982 and 1994. Overall, this total represents 19% more cancer than would have been expected over this 12-year period. In absolute terms, this is 17 excess cases. The three most common types of cancer diagnosed were lymphomas (n = 26), leukemias (n = 23), and central nervous system tumors (n = 17) (other cancers were renal 8, soft tissue 10, carcinoma 9, other 7). The observed elevation in total cancer was mostly due to elevations in lymphomas and leukemia among mid-Cape children, and to a lesser extent leukemia among lower-Cape residents (see Table 9-5). Childhood cancer did not appear to concentrate in specific areas within towns or within specific time periods over the 12 years of data accumulation. In addition, the types of cancer most commonly observed in the ages of diagnosis generally followed patterns observed elsewhere in Massachusetts. These results do not suggest common environmental factors being responsible for the observed excess childhood cancer.

Birth Defects

In 1996, the Centers for Diease Conrol and Prevention awarded five years of funding to the Massachusetts Department of Public Health to establish the Massachusetts Center for Birth Defects Research and Prevention. An active birth defects surveillance system involves trained personnel who validate passive reports of birth defects cases to the Department, and actively seek cases in hospitals and other health facilities. This approach provides more complete ascertainment of cases, more accurate data on cases, and more rapid reporting of cases. The center collected data on birth defects in the eastern part of Massachusetts from October 1997 to October 1998. Since October 1998, the surveillance program has been collecting cases statewide from the 53 birth hospitals and Children's Hospital, Boston. The report of 1999 statewide data represents the first full year of surveillance data (MDPH 2001).

All cases represent diagnoses before the infant reached one year of age. The

TABLE 9-5 Standardized Incidence Rates, Childhood Cancer, Cape Cod, 1982-1994

	Observed	Expected	SIR (95% CI)
All of Cape Cod			
Leukemia	23	20.5	112 (71-169)
Lymphoma	26	13.3	195 (127-286)
CNS	17	21.3	80 (46-128)
Upper Cape Cod			
Leukemia	12	13.0	92 (48-192)
Lymphoma	11	8.3	132 (66-236)
CNS	10	13.5	74 (35-136)
Mid Cape Cod			
Leukemia	7	5.9	120 (48-246)
Lymphoma	13	3.9	332 (177-568)
CNS	7	6.1	114 (46-235)
Lower Cape Cod			
Leukemia	4	1.6	Nc[a]
Lymphoma	2	1.1	Nc
CNS	0	1.7	Nc

[a]Not calculated.

infant or fetus must have a structural defect that met diagnostic criteria, and the infant was live born, or the fetus was stillborn with gestational age greater than or equal to 20 weeks or with a weight of at least 350 grams.

Overall for Massachusetts in 1999, 1.1% of births in the state had one or more birth defects. The rate of birth defects per 10,000 births was 108.2 (95% confidence interval 101.2 to 115.5). This was based on a total of 875 diagnosed and reported birth defects. The most common defects among live births and stillbirths in 1999 were atrial septal defect (95 cases), ventricular septal defect (71 cases), cleft lip with and without cleft palate (69 cases), trisomy 21 (Down syndrome, 67 cases), hypospadias (2nd and 3rd degree, 49 cases), cleft palate without cleft lip (44 cases), obstructive genitourinary defect (42 cases), pulmonary valve atresia and stenosis (34 cases), coarctation of the aorta (25 cases), and endocardial cushion defects (25 cases).

When the rate is evaluated by region in the state, the lowest rate is reported for the southeast, a region that includes the Cape and islands and extends north to encompass a total population generating 15,000 births. The 130 cases of birth defects diagnosed in 1999 represents a rate of 86.1 per 10,000 births (95% confidence interval 71.9 to 101.5). The apparently lower rate may be because of some births being delivered in Rhode Island where Massachusetts surveillance was not conducted. The limited number of birth defects each year precludes firm conclusions and requires accumulation of additional years of surveillance data to generate stable estimates of rates.

SAMPLE SIZE AND STATISTICAL POWER TO INVESTIGATE HEALTH RISKS OF THE PAVE PAWS RADAR

Introduction

When investigating the occurrence of disease within a defined population, it is important to recognize the inherent limitations that may be encountered when making a determination of whether or not there is an excess in the number of expected cases. Disease rarely occurs in a uniform manner within the general population. That is, disease within a defined population can, by random chance, vary in occurrence according to time and place. Thus, in order to evaluate if disease is occurring in excess, one must take into consideration the random variation that can occur. This is done by statistically considering the probability that a given observation may occur by chance alone. Conversely, one is often faced with the question of what level of excess can be ruled out with some reasonable level of confidence.

The concept of statistical power refers to the ability to detect a given level of excess within a defined population. Calculation of statistical power requires assumptions and/or estimates of the (1) expected incidence of disease, (2) proportion of the population exposed to the factors of interest, (3) size and composition of the population, and (4) level of risk one would like to detect/rule out. Consider the following example. Prior to embarking on a study to determine if there is twofold or greater occurrence of a specific disease among the residents of a designated city, a researcher calculated the statistical power to be 0.75. This means that if a twofold excess truly does exist the researchers will have a 0.75 (i.e., 3 out of 4) chance of detecting the twofold difference.

Power Calculations

Using cancer as an example, the following calculations have been made for the upper Cape Cod population assuming (1) cancer rates for the United States obtained from the Surveillance, Epidemiology and End Results Program, (2) 12% of the residents in upper Cape Cod reside in a location that results in exposure to the sidelobes of the PAVE PAWS radar beam, (3) the population size and composition of upper Cape Cod is that obtained for the 2000 U.S. Census (see Figure 9-1) and would be observed over a 10-year period, and (4) alpha = 0.05 for a two-sided test. The power calculations have been made for a spectrum of risk, ranging from 1.1 (i.e., 10% increase) to 2.5 (250% increase).

However, it is not likely that any single-source exposure would etiologically

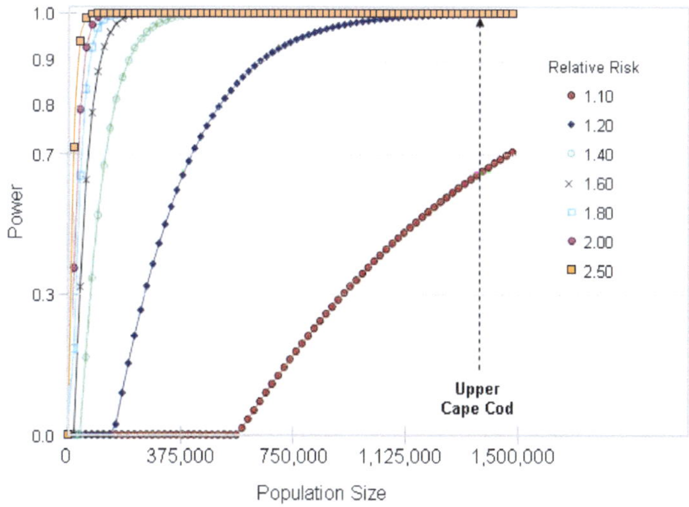

FIGURE 9-1 Power to detect total cancer (excluding non-melanoma skin cancer) occurrence.

be associated with an increase in the broad spectrum of malignancies seen in the general population. Rather, it is more likely that if the PAVE PAWS radar is associated with an increase in cancer risk, there will be some level of specificity as to the types/forms of malignancy. Therefore, it is important to consider statistical power within a cancer-specific context.

Summarized in Table 9-6 is the diagnosis-specific power calculation for the upper Cape Cod population. Highlighted are those cancer-specific groups where power would be considered sufficient to detect a modest level of risk (i.e., less than twofold).

Concerns have been raised regarding the potential excess of malignancies among the pediatric population in the upper Cape. Since the incidence of cancer in children under the age of 15 years is relatively low, compared to rates among the adult populations, it is not surprising that the power to detect anything except extremely large (i.e. >100-fold) risks does not exist. Even when considering all diagnoses combined, there would be 80% power to detect a 9-fold (900% increase). As with adult malignancies, it is unlikely that a single exposure (e.g., radar) would increase the occurrence of multiple types of malignancy. Accordingly, it is not possible to study increases of specific pediatric cancers in the upper Cape Cod in order to rule out risks less than 100-fold in excess.

For the residents of Cape Cod, it is unclear what minimal level of risk would provide sufficient assurance that the PAVE PAWS facility is not adversely impacting health. If the answer is 99% assurance that there is not an increase of

TABLE 9-6 Minimum Relative Risk Detectable in Population Followed for 10 Years, Assuming 12.5% of the Population Exposed, and Background SEER Cancer Incidence Rates 1975-2001 (All Ages), Alpha = 0.05 Two-Sided Test

	Total population (25,300)		Males (13,000)		Females (12,300)	
	Power =. 8	Power = .9	Power =. 8	Power =. 9	Power= . 8	Power = .9
Any Cancer	1.12	1.14	1.18	1.21	1.17	1.20
Colon & Rectum	1.42	1.49	1.69	1.84	1.65	1.78
Lung & Bronchus	1.38	1.45	1.62	1.75	1.59	1.70
Breast	1.34	1.40	>100.00	>100.00	1.35	1.41
Corpus Uteri	—	—	—	—	3.11	4.01
Ovarian	2.89	3.64	—	—	5.09	8.30
Prostate	—	—	1.55	1.65	—	—
Testicular	—	—	78.52	>100.00	—	—
Bladder	1.81	2.00	2.49	2.99	2.37	2.81
Kidney & Renal Pelvis	2.33	2.73	3.81	5.34	3.53	4.78
Hodgkin Lymphoma	3.80	22.18	>100.00	>100.00	62.79	>100.00
Non-Hodgkin Lymphoma	1.85	2.06	2.59	3.13	2.46	2.93
Lymphoid Leukemia	3.94	5.62	10.67	33.84	9.00	23.22
Myeloid Leukemia	3.53	4.78	8.43	20.33	7.28	15.33
Myeloma	3.63	4.99	8.96	23.01	7.69	16.99
Pancreas	2.34	2.75	3.84	5.41	3.55	4.83
Liver	3.94	5.62	10.67	33.84	9.00	23.22
Soft Tissue	7.81	17.49	70.25	>100.00	43.74	>100.00
Bone & Joint	>100.00	>100.00	>100.00	>100.00	>100.00	>100.00

greater than 1%, then research would not be able to satisfactorily address such concerns. Using cancer as an example and assuming 10 years of observation of the population, one could reasonably anticipate that well-designed and well-conducted research, either prospective or retrospective in design, could address moderately low levels of risk (i.e., between 1.4 and 2.0) for some of the more common cancers (i.e., colo-rectal, lung, breast, prostate, bladder, and non-Hodgkin's lymphoma). For other diagnoses the ability to address less than a twofold risk would require extended length of observation of exposed residents.

STATISTICAL ANALYSES

Introduction

To characterize challenges inherent in the epidemiological assessment of the health effects associated with exposure to the PAVE PAWS radar, we conducted a geographical correlation study within the upper Cape Cod region surrounding the Massachusetts Military Reservation (MMR) where PAVE PAWS is located. Our statistical analysis focused mainly on the 30 census tracts in five towns shown in Map 1, and it was organized as follows. First, we conducted an exposure-

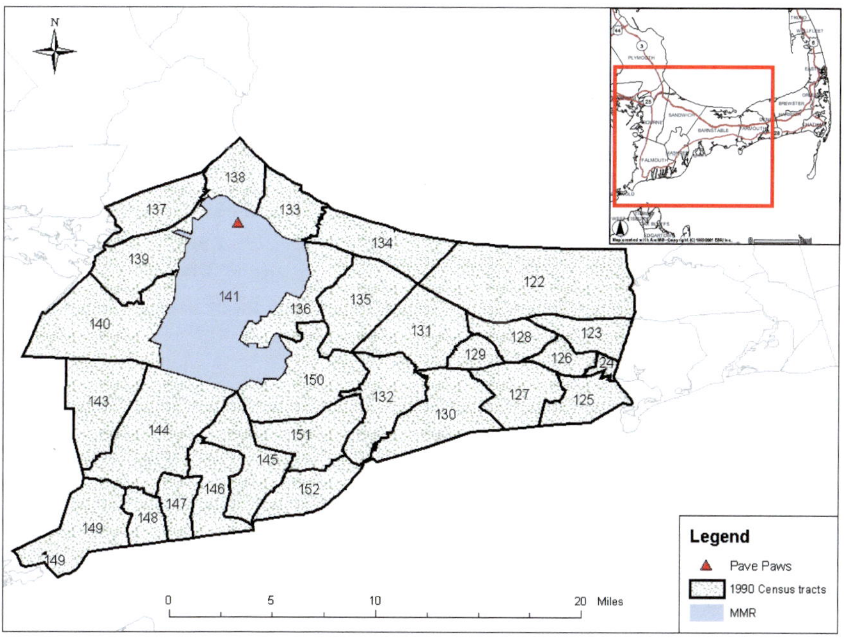

MAP 1 Location of 30 census tracts in upper Cape Cod region studied in statistical analysis.

assessment study where we estimated: (a) the percent of the population in the Cape that is exposed to the PAVE PAWS radar by age and gender; and (b) the power-density exposure (average and peak) from three different sources at each census tract weighted by population density. Second, we performed a geographic correlation study to estimate associations between Standard Incidence Ratios (SIR) and estimated exposure to PAVE PAWS adjusted by socio-economic status and other spatially correlated factors.

EXPOSURE ASSESSMENT STUDY

Percent of Population Exposed to PAVE PAWS Radar Beam

We first estimated the geographical areas in the Cape that had a non-zero exposure to the PAVE PAWS radar beam. Those areas were identified by taking into account: (a) that exposure occurs only in the area with direct line of sight to the radar, and (b) the operating triangle of the radar (an area has direct line of sight to the PAVE PAWS radar only if there is no higher ground in between). Because the main beam of the radar ends at 0.1 degree above the horizontal, these areas are exposed to the power radiated primarily by the sidelobes of the radar, which have much lower power densities than the main beam. We determined regions with line of sight to the PAVE PAWS radar by using a digital elevation map (Map 2) from the GeoComm International corporation.[4]

We then further refined the calculations by taking into account the total height of the radar (105 feet) and the PAVE PAWS radar scanning angle (347 degrees to 227 degrees). The regions with non-zero exposure are shown in Map 3. Map 4 shows the estimates areas with non-zero exposures as provided bo use by MITRE.

To estimate the percent of the population exposed to the PAVE PAWS radar beam we proceeded as follows. First, we used the 1990 and the 2000 census-bureau data to determine the population size by age and gender at each location. Because of the small sample size within each census block, we aggregated the population density by census-block group (Map 5).

Within each census-block group, we estimated the number of persons exposed to the radar by multiplying the population in that census-block group by the proportion of area within that census block group with non-zero exposure. We then summed the estimated number of persons exposed across the 117 census-block groups (102 census-block groups in year 2000), and divided the sum by the total population. Overall we estimated that in 1990, 11.8% of the population living in upper Cape Cod was exposed to the PAVE PAWS radar beam. Percentages of the total population exposed to the PAVE PAWS radar beam by age and gender are presented in Table 9-7.

[4]http://gisdatadepot.com/dem May 2004.

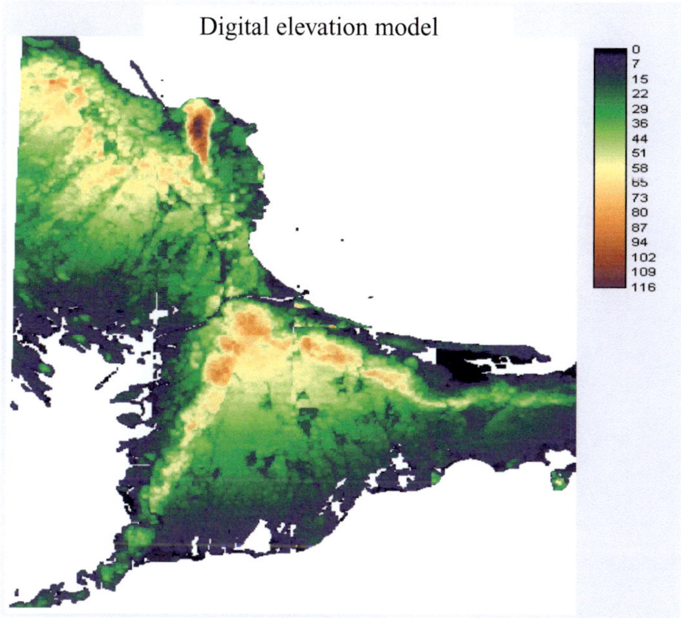

MAP 2 Digital elevation model. Data from GeoComm International Corporation.

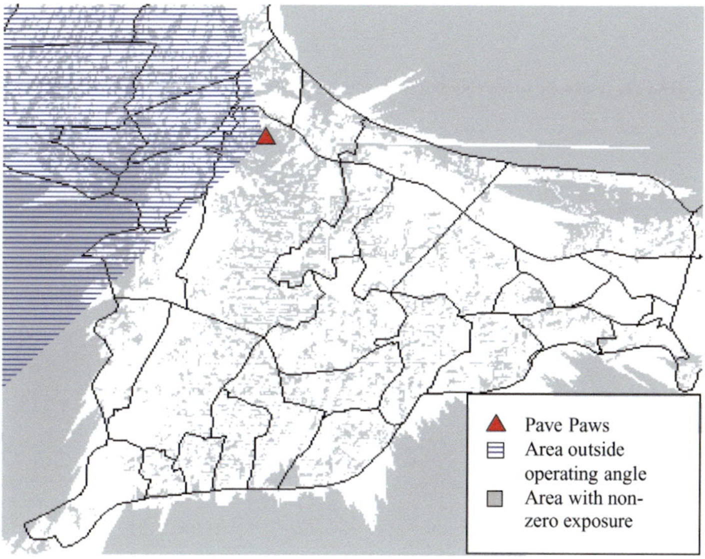

MAP 3 Shaded areas over land indicate non-zero exposures.

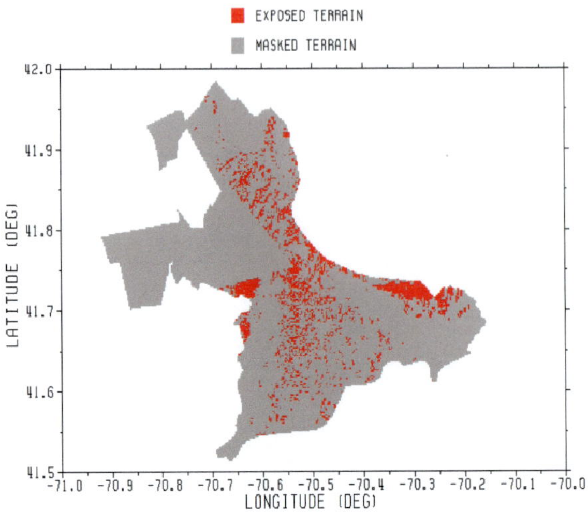

MAP 4 Estimated areas with non-zero exposure as provided by MITRE.

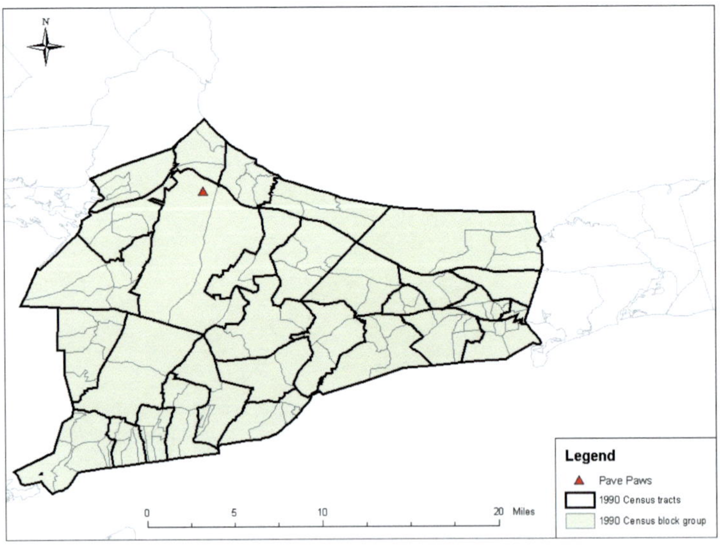

MAP 5 117 Census block groups within 30 Census tracts in 1990.

TABLE 9-7 Estimated Proportion of Population Living Under the Line of Sight of PAVE PAWS Radar

	1990 Census				2000 Census			
	Male		Female		Male		Female	
Age Group	Population Size	Proportion Exposed	Population Size	Proportion Exposed	Population Size	Proportion Exposed	Population Size	Froportion Exposed
All ages	51,759	11.8%	56,481	11.8%	63,213	12.4%	69,071	12.4%
0-4	3,989	12.3%	3,774	12.5%	3,736	13.5%	3,505	13.8%
5-14	6,959	12.5%	6,757	12.7%	9,216	13.4%	8,819	13.2%
15-24	6,478	10.3%	5,970	11.1%	6,649	11.0%	5,949	11.6%
25-34	8,238	11.6%	8,841	11.7%	6,507	12.1%	7,128	12.3%
35-44	8,226	12.5%	8,653	12.2%	10,282	12.6%	11,259	12.5%
45-54	5,010	12.0%	5,469	11.8%	9,283	12.4%	10,149	12.5%
55-64	4,851	11.8%	5,754	11.3%	6,593	12.3%	7,805	12.3%
65-74	5,113	11.8%	6,227	11.7%	6,302	12.3%	7,323	11.9%
75-84	2,348	11.8%	3,617	10.7%	3,751	12.0%	5,161	11.9%
85+	547	10.3%	1,419	9.7%	894	11.2%	1,973	11.3%

There was little difference in exposure between each age and gender group. Based on 2000 census data, the estimated proportion of population living in the area exposed to the PAVE PAWS radar beam increased slightly to 12.4%. The changes in population distribution in the past decade had little effect on the proportion of population living under the line of sight from the PAVE PAWS radar.

Average and Peak Power-Density Assessment

Because cancer-incidence data was made available to us only at census-tract level, to carry out a geographic correlation study and to estimate association between cancer incidence and exposure to the PAVE PAWS radar beam, we needed to estimate average and peak power density at each census tract. Three sets of power-density data were used in this analysis. The first set was calculated based on two simple assumptions. The first assumption was that only the area with direct line of sight to the radar and under the radar's operation angle would be exposed to the radar beam. The second assumption was that power density decreased as the distance from the radar increased. This relationship was estimated by fitting a power curve between the average power-density obtained from_the MITRE Technical Report (MITRE 2000) and the distance to the radar. The fitted curve is equal to average power density ($\mu W/cm^2$) = $8972*distance^{-1.325}$(meters).

With these two assumptions, we calculated both average and peak power-density at locations 180 meters apart within the study region (Map 6).

The committee is aware that these are very simplistic assumptions. Nonetheless, this first dataset provided a good basis for validating datasets from other sources. In this report, we refer to this estimated exposure to the PAVE PAWS beam as "simple."

The second estimated power-density dataset was provided by MITRE Corporation at locations 200 meters apart (Map 7).

Detailed descriptions of the PAVE PAWS radar power-density calculations can be found in the MITRE technical report (MITRE 2000). The third dataset was obtained from Broadcast Signal Lab (BSL) (Map 8).

To estimate the radiofrequency exposure from the PAVE PAWS radar in each census-block group on Cape Cod (BSL 2004), BSL utilized "the ComStudy propagation modeling software" in conjunction with an analytical model of the PAVE PAWS antenna supplied by the MITRE Corporation. We obtained power-density estimates at locations 140 meters apart in the east-west direction and 188 meters apart in the north-south direction (Map 8).

We then estimated exposure at each census tract by weighting power-density values in $\mu W/cm^2$ at each location by the corresponding population density at the census block (Map 9).

Population density was calculated by dividing the number of individuals living in the census block in 2000 by total square miles of dry land. The 2000 census data were used because the census bureau did not provide 1990 census informa-

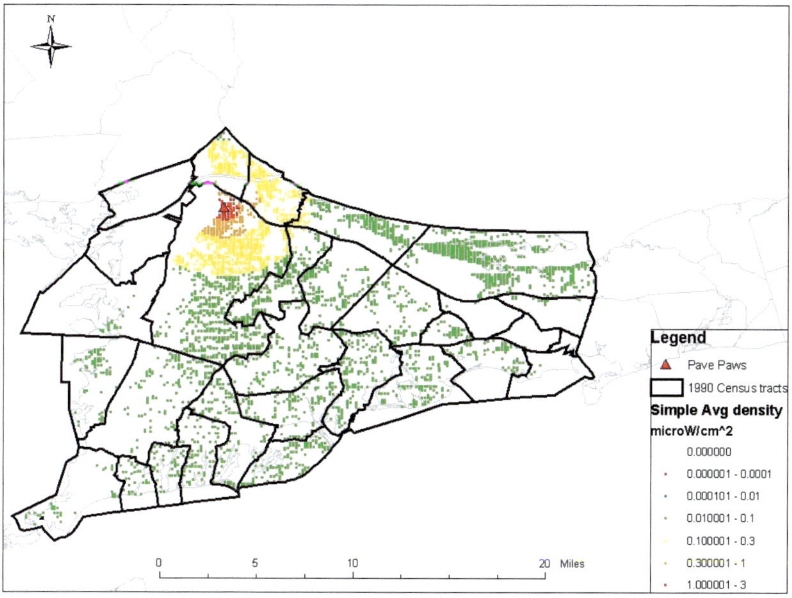

MAP 6 Simple average power densities.

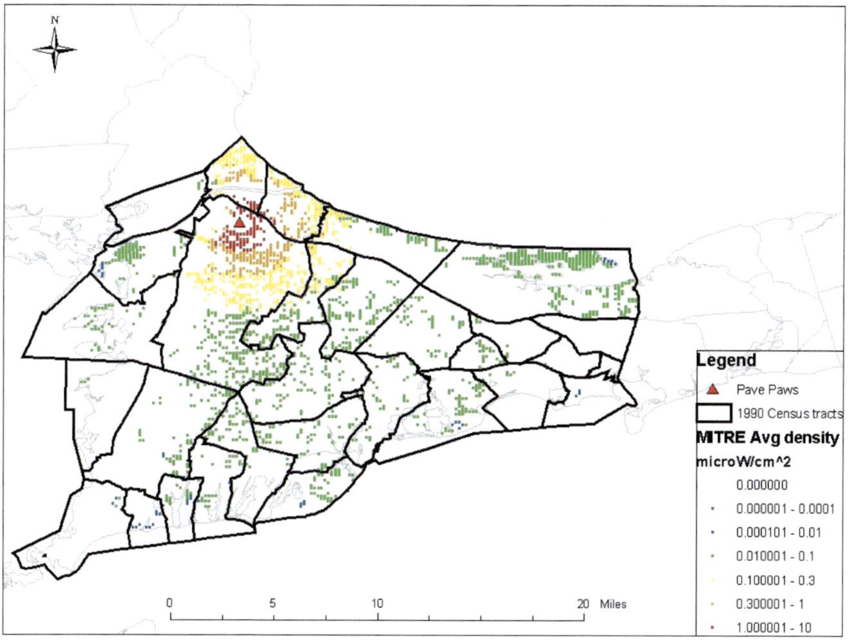

MAP 7 MITRE average power densities.

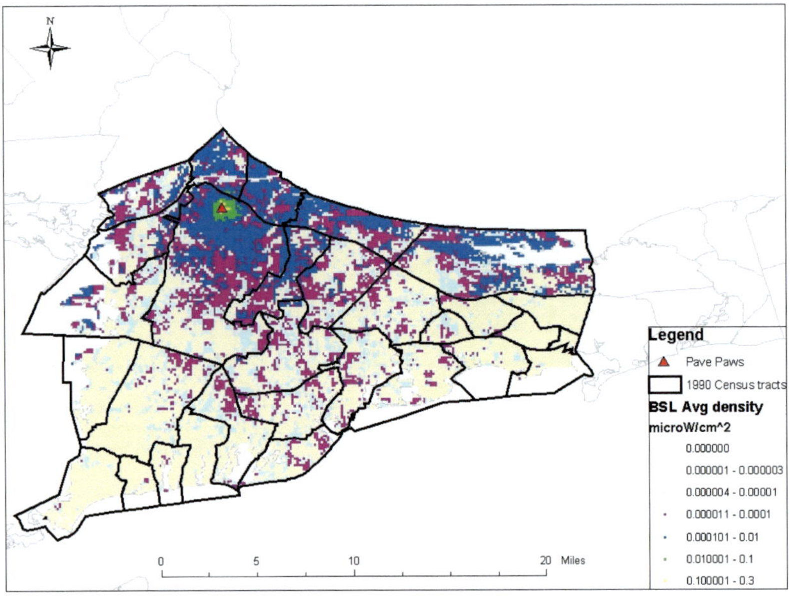

MAP 8 BSL average power densities.

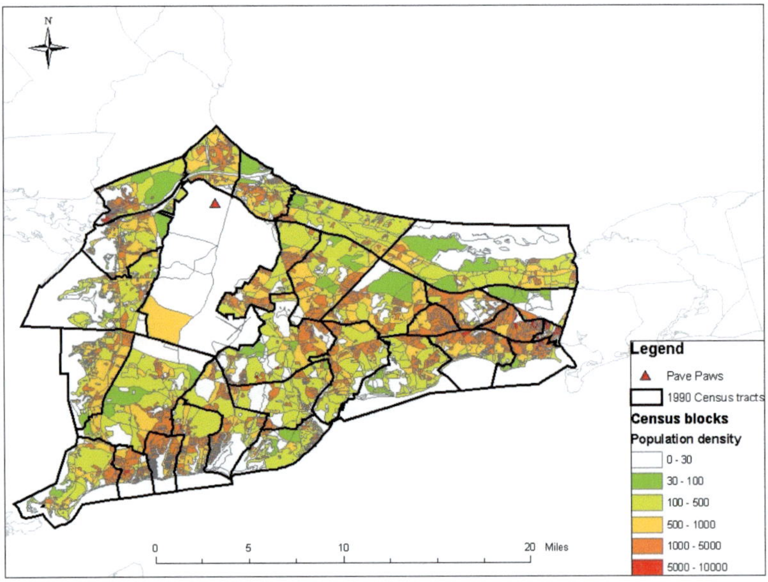

MAP 9 Population densities at census block level.

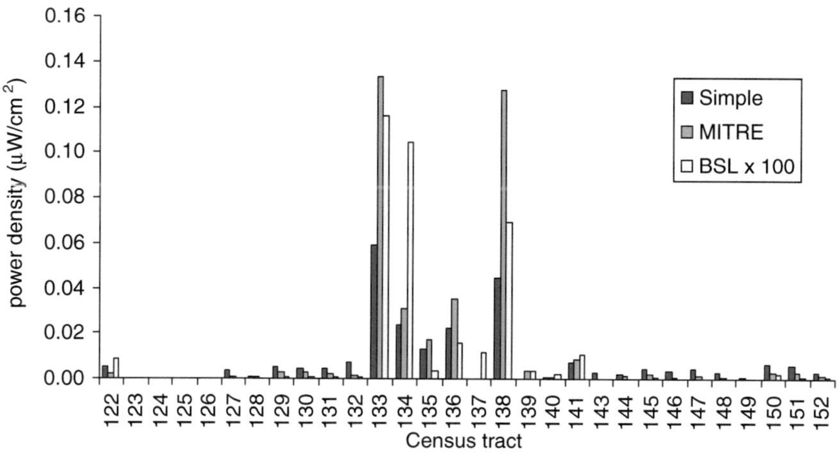

FIGURE 9-2 Average density estimate comparison.

tion at the census-block level. We then aggregated the exposure estimates at each location to obtain average and peak power-density exposures in each census tract in 1990. The resulting average power-density exposure estimates at the census-tract level from the three data sources are shown in Figure 9-2. (Note that results shown for BSL are magnified by a factor of 100.)

As expected, since the "simple" exposure estimates partly depended on MITRE power-density estimates, these two set of results were very similar. Although BSL predicted non-zero exposure in almost the whole study region, its resulting estimates were generally lower than those from MITRE. Most census tracts having high exposure estimates from the MITRE dataset also had relatively high estimates from the BSL dataset. For census tract 137, the difference was due to a large proportion of its area being situated outside the PAVE PAWS operating angle and that BSL estimates took into account reflection of the radar signal in this region. It is important to notice that the absolute magnitude of the density estimates would not affect the statistical significance of the variable of interest in our geographic-correlation study.

Socio-economic Factor Adjustment

In the period of 1986-1994, total-cancer incidence was 11% higher in upper Cape Cod compared to the state of Massachusetts (SIR 95% CI: 108-113; see Table 9-2). This elevated SIR might be caused by three potential confounding factors: age distribution, smoking behavior, and socio-economic status (SES). Confounding due to age distribution can be controlled in the calculation of the expected cancer-incidence counts. In the upper Cape Cod cancer-incidence review, the expected counts were calculated using 10 age categories in lieu of 6 age categories used by Massachusetts Cancer Registry in its 1997 report. These additional age groups provided extra precision in the calculation of the expected cancer-incidence count, addressing the concern of confounding by age. Since smoking has been identified as a risk factor of multiple cancers, it is important to find out if differences in smoking behavior exists between the population in upper Cape Cod and in the state of Massachusetts. Results from Massachusetts Behavioral Risk Factor Surveillance System (BRFSS) for the period 1999 to 2001 estimated similar proportions of current smokers with 19% and 20% in upper Cape Cod and the state of Massachusetts, respectively. The small sample size in this study (304 in upper Cape Cod) provides little statistical power to demonstrate any difference in the estimates. In the Health Risks and Preventive Behaviors report that combined BRFFS results from 1994 to 1999, the estimate of current smokers in the Cape and islands was 18.3% (95% CI 14.8%-21.7%) compared to 21.2% (95% CI 20.4%-21.9%) in Massachusetts. This difference is not statistically significant. The last concern is the difference in SES between the population in upper Cape Cod and in the state of Massachusetts that may change the expected number of cancer incidences. For example, if one type of cancer has a higher detection rate in a population with high socio-economic status due to better access to health care and one supposes that a wealthy population lives in upper Cape Cod, then one should expect to see more cancer diagnosed in upper Cape Cod. Failing to adjust for this higher than expected number of cancer incidences will over-estimate the SIR. To assess the potential confounding due to SES disparity, we examined the percentage of population older than 18 years living below the line of poverty from the 1990 census data.

Upper Cape Cod has 6.5% of its population living below the poverty line compared to 8.3% in Massachusetts (see Figure 9-3). This difference required the

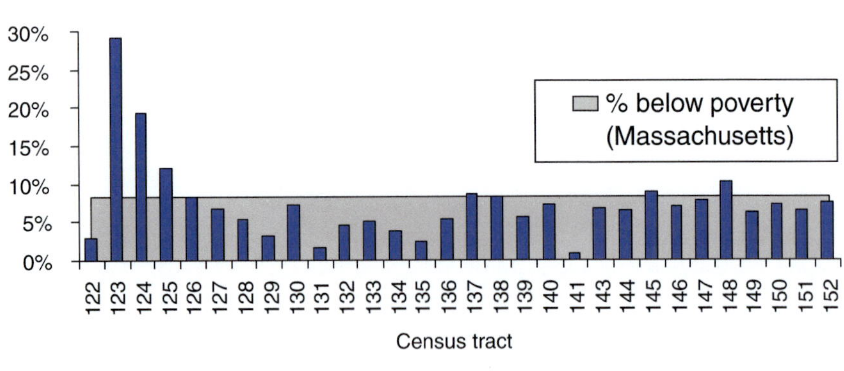

FIGURE 9-3 Percent population (18 years +) living below line of poverty 1990.

Model 1

$$SIR^{c}_{SESadjusted} = O^{c} / E^{c} \quad \text{where } E^{c} = \sum_{i}\sum_{j}\sum_{k} n^{c}_{ijk} * r_{ijk}$$

O^c : observed number of cancer incidence in census tract c
E^c : expected number of cancer incidence in census tract c
n^{c}_{ijk} : number of person in age group i, gender j, poverty level k, census trace c
r^{c}_{ijk} : cancer incidence rate for age group i, gender j, poverty level k in Massachusetts

adjustment of the cancer SIR by socio-economic status in addition to the standard adjustment by age and gender.

Model 1 describes the indirect adjustment made to obtain SIR further adjusted by socio-economic status. One important data requirement is the knowledge of the cancer incidence rate for a particular age, gender, and poverty group in Massachusetts. However, limited SES-related information is collected by the Massachusetts Cancer Registry.

Because of data unavailability for indirect SES adjustment on cancer-incidence ratios, we explored the potential association of SES and cancer SIR using data from the "Upper Cape Cod Cancer Incidence Review, 1986-1994" and SES data from the Census Bureau. Since 62% of all cancer incidences in upper Cape Cod during that period were one of the four major cancer types (colo-rectal, breast, lung, prostate cancer), we used the SIR of these four major cancer types as response variable, and the percentage of population older than 18 living below the poverty line as a measure of socio-economic status. These data were fitted using a log-linear model (Model 2) taking into account unobserved explanatory variables that are assumed to be spatially correlated (MDPH 2002). The parameter γ in this model represents the association between SIR and SES, namely how SIR varies with respect to changes in SES. This association will be presented as per-

Model 2

$O_i|\mu_i \sim \text{Poisson}(\mu_i)$ $\quad$ i = census tracts
$\log \mu_i = \log E_i + \alpha_0 + \gamma^* SES_i + b_i$
α_0 : overall log relative risk
γ : log relative risk of cancer incidence associated with SES
b_i : spatial correlation

TABLE 9-8 Association Between Socio-economic Status and Cancer Incidence Ratio

Cancer Types	% Change in Relative Risk per 1% Increase in Population Below Poverty	95% C.I.
Colo-rectal cancer	2.48	0.54 - 4.57
Breast cancer	–1.04	–2.72 - 0.59
Prostate cancer	–0.90	–3.79 - 1.90
Lung cancer	1.49	–0.78 - 3.53

cent change in relative risk of cancer incidence for each percent increase in population above age 18 living below the line of poverty.

All models were implemented in WinBUGS 1.4 statistical programming language. We employed a standard procedure reporting small coefficient estimates γ and α_1 as a percent change in relative risk per unit change in SES and exposure, respectively. Besides colo-rectal cancer, our analyses of cancer incidence on upper Cape Cod demonstrated little association between socioeconomic factors and the standardized incidence ratio of three major cancer types (see Table 9-8).

In the case of colo-rectal cancer, data showed 2.5% increase in relative risk (95% CI: 0.5-4.6%) per 1% additional population living below poverty. This suggested that colo-rectal cancer SIR could be understated for higher-income census tracts with higher-income levels.

Furthermore, if these higher-income census tracts had higher exposure, then we might underestimate the association between exposure and SIR. Plotting SES by exposure (Figure 9-4) revealed no strong relationship between these two variables; hence, results from our geographic correlation study should not be sensitive to the SES adjustment.

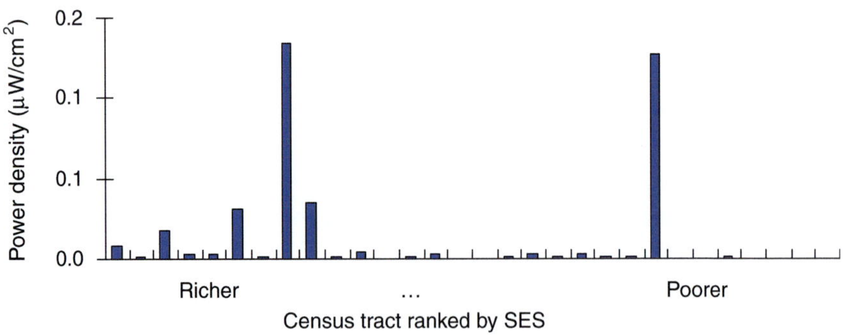

FIGURE 9-4 Census tract MITRE average-density exposure by SES ranking,

GEOGRAPHIC CORRELATION STUDY

To measure the association between the estimated PAVE PAWS power-density exposure (average and peak) and cancer incidence at census-tract level, we fitted a log-linear model taking into account spatial correlation for unobserved variables. In addition, we controlled for SES by introducing the percentage of population over 18 years of age living below the poverty level in our model (Model 3). For each of the four major cancer types, the cancer incidence data revealed little evidence of association with our estimated exposure and we determined that controlling for SES did not lead to qualitative changes in the relative rate estimates (Figure 9-5).

Model 3

$O_i|\mu_i \sim \text{Poisson}(\mu_i)$ i = census tracts

$\log \mu_i = \log E_i + \alpha_0 + \alpha_1 * Exposure_i + \gamma * SES_i + b_i$

α_0 : overall log relative risk

α_1 : log relative risk of cancer incidence associated with PAVE PAWS density exposure

γ : log relative risk of cancer incidence associated with SES

b_i : spatial correlation

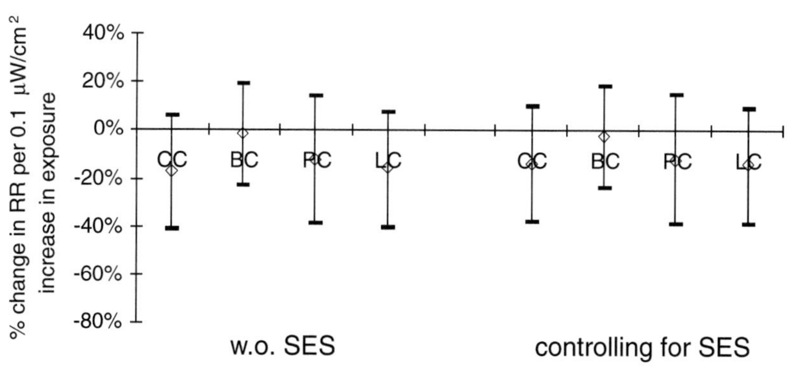

FIGURE 9-5 Association between MITRE average power-density exposure and cancer incidence with/without controlling for SES. CC—colo-rectal cancer, BC—breast cancer, PC—prostate cancer, LC—lung cancer.

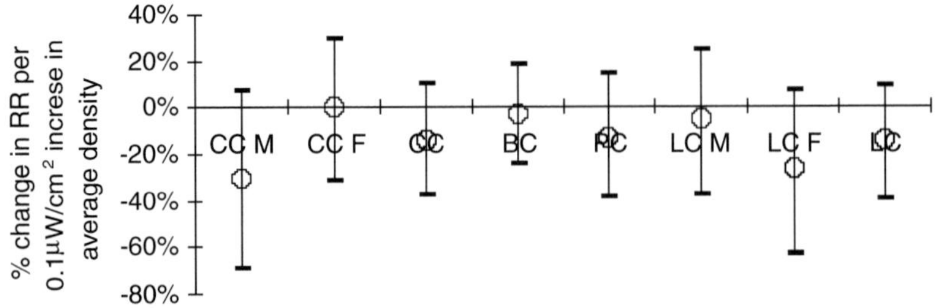

FIGURE 9-6 Association between MITRE average density exposure and cancer incidence. CC—colo-rectal cancer, BC—breast cancer, PC—prostate cancer, LC—lung cancer. M = male, F = female.

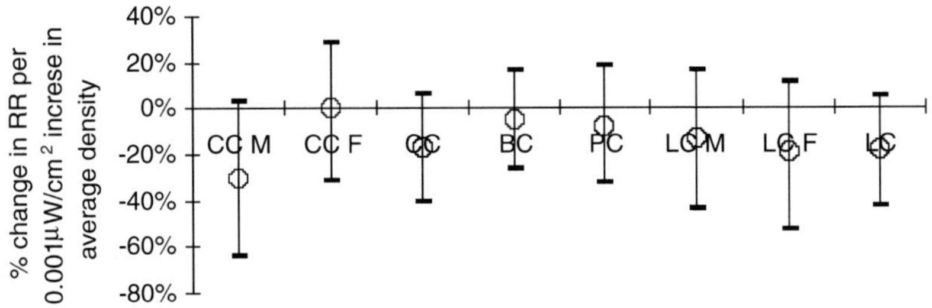

FIGURE 9-7 Association between BSL average density exposure and cancer incidence. CC— colo-rectal cancer, BC—breast cancer, PC—prostate cancer, LC—lung cancer. M = male, F = female.

Based on Model 3, we found no statistically significant association between MITRE average power-density exposure estimates and any of the four major cancer types (Figure 9-6). We also conducted a sensitivity analysis by fitting the same model using BSL average power-density exposure estimates (Figure 9-7) and MITRE peak power-density exposure estimates (Figure 9-8). We found that our result was not sensitive to the choice of exposure estimates.

Due to the large power-density estimates at census tracts 133-6 and 138, results in Figures 9-6 to 9-8 might be highly influenced by these few census tracts. We, therefore, refitted the statistical model by using BSL average power-density expressed in decibels with respect to microwatts per centimeter square ($dB\mu W/cm^2$) where exposure in $dB\mu W/cm^2 = 10*\log_{10}$ exposure in $\mu W/cm^2$. Similar to the spatial correlation analysis conducted in the $\mu W/cm^2$ scale, after controlling for SES, the associations between exposure and cancer incidence were not statistically significant for all four major cancers (Figure 9-9).

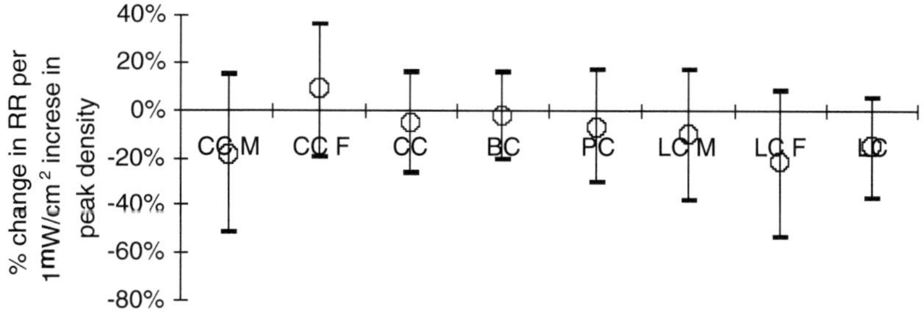

FIGURE 9-8 Association between MITRE peak-density exposure and cancer incidence. CC—colo-rectal cancer, BC—breast cancer, PC—prostate cancer, LC—lung cancer. M = male, F = female.

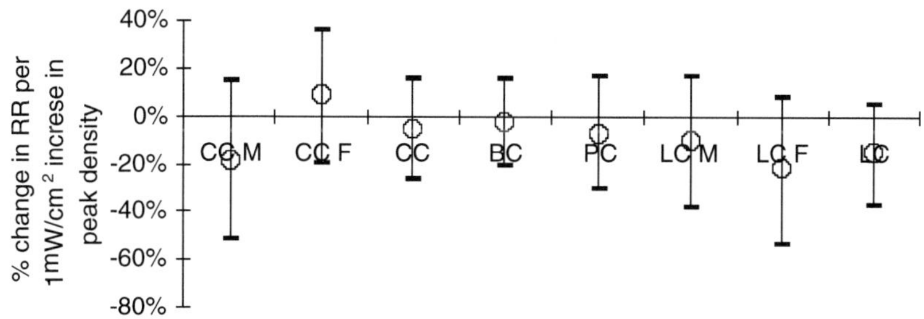

FIGURE 9-9 Association between BSL average density exposure (decibel) and cancer incidence. CC—colo-rectal cancer, BC—breast cancer, PC—prostate cancer, LC—lung cancer. M = male, F = female.

Geographic Correlation Study Summary

To combine evidence of all 24-cancer types, we fit a hierarchical model as specified in Model 4. Neither of the two exposure estimates showed statistically significant association with cancer incidence between 1986 and 1994. For peak-density exposure, we estimated a 6.5% decrease in relative risk (95% CI: –14.2 - 1.2%) per 1 $\mu W/cm^2$ increase in the level of peak exposure. In addition, we estimated a 6.6% decrease in relative risk (95% CI: –11.8 - 0.1%) per 0.1 $\mu W/cm^2$ increase in average-density exposure.

Findings of the National Research Council committee's geographical correlation study by census tract of exposure to PAVE PAWS radar and cancer incidence should be interpreted with caution. First, because of the lack of sufficient health-related statistics and confounders at a geographical resolution smaller than a census tract, our results could be affected by ecological and confounding bias. In addition, as in most environmental risk-assessment studies, our census-tract

Model 4

$O_i^k \mid \mu_i^k \sim \text{Poisson}\left(\mu_i^k\right)$ $i =$ census tracts, $k =$ cancer types

$\log \mu_i^k = \log E_i^k + \alpha_0^k + \alpha_1^k * Exposure_i + b_i^k$

$\alpha_0^k \sim N(\alpha_0, \tau_o)$

$\alpha_1^k \sim N(\alpha_1, \tau_1)$

α_1: overall log relative risk of cancer incidence associated with PAVE PAWS density exposure

exposure estimates could be affected by exposure-measurement error. In fact, census-tract exposure estimates are residential and might not represent the actual personal exposures. Epidemiological studies that use health-related statistics (mortality, morbidity, and pregnancy outcomes) and potential confounders (smoking, pollution, SES) at a smaller level of spatial aggregation (i.e., census-block group) and that take into account daily activity patterns in the exposure assessment would provide a more precise and comprehensive assessment of the potential health effects associated with an exposure to the PAVE PAWS radar.

SUMMARY

1. Evaluation of previously reported data demonstrates that:

- In general, there is an absence or lack of sufficient health-related statistics available for Cape Cod residents to allow for a detailed descriptive analysis within smaller and more specific subdivisions such as counties, towns, census tract, and census block. Population-based data are available only for outcomes relating to cancer, birth defects, birth weight, and premature mortality, although the size of the population within geopolitical areas of interest does not always allow for a sufficient number of outcomes to allow adequate evaluation.
- The observed elevated cancer-incidence rates among residents of Cape Cod have not been adequately explained through subsequent investigations.
- While previous case-control and ecologic correlation studies have not provided compelling evidence for elevated cancer risk associated with exposure to the PAVE PAWS radar, it cannot be concluded from these studies that the PAVE PAWS radar is not associated with cancer risk. The previous studies have limitations relating to cancer-specific sample size and detailed exposure assessment.
- Because of the limited population of upper Cape Cod, it will not be

possible to conduct cancer-specific research studies with statistical power to rule out small or moderate levels of increased risk related to exposure to the PAVE PAWS radar beam.

- The International Epidemiology Institute has proposed a study at the county level using counties west of PAVE PAWS (out of PAVE PAWS operating angle) as a control. This study proposes to evaluate health outcomes before and after the startup of the PAVE PAWS radar, which should strengthen any association, or lack thereof, of health effect with the PAVE PAWS radar.

2. Geographical correlation by census tract of exposure to PAVE PAWS radar and cancer incidence (analysis conducted by this NRC committee):

- Using available information on population density, topography, and direction of the PAVE PAWS radar beam it is estimated that about 16,400 people in Cape Cod reside in the sidelobes of the main PAVE PAWS beam.
- Analysis of the occurrence of specific cancers, including colorectal, breast (female), prostate, and lung, did not identify any significant association with estimated peak or average power-density exposure to the PAVE PAWS beam.
- Statistical adjustment for socioeconomic status, utilizing the percent of population below the poverty level, had little impact on the observed correlations. The lack of census-tract level information on confounding factors such as smoking and diet remains a limitation of our geographical correlation study. However, for a confounding factor to have significant effect on the results, it needs to be strongly related to the cancer incidence and inversely correlated with exposure estimates.

REFERENCES

Aschengrau, A., C. Paulu, and D. Ozonoff. 1998. Tetrachloroethylene-contaminated drinking water and the risk of breast cancer. Environ Health Persp 106 (Suppl 4):947-953.

Aschengrau, A., S. Rogers, and D. Ozonoff. 2003. Perchloroethylene-contaminated drinking water and the risk of breast cancer: additional results from Cape Cod, Massachusetts. Environ Health Persp 111:167-73.

BSL (Broadcast Signal Lab). 2004. Draft Final Test Report on a Survey of Radio Frequency Energy Field Emissions from the Cape Cod Air Force Station PAVEPAWS Radar Facility, Broadcast Signal Lab, Medfield, MA.

CCCR (Cape Cod Commission Reporter). April 17, 2003. Vol 13, no 4.

Colditz, G. 1998. Relationship between estrogen levels, use of hormone replacement therapy, and breast cancer. J Natl Cancer I 90:814-823.

Coogan, P.F. and A. Aschengrau. 1998. Exposure to power frequency magnetic fields and risk of breast cancer in the Upper Cape Cod Cancer Incidence Study. Arch Environ Health 53:359-367.

Coogan, P.F. and A. Aschengrau. 1999. Occupational physical activity and breast cancer risk in the Upper Cape Cod Cancer Incidence Study. Am J Ind Med 36(2):279-285.

Eyles, J. and S. Birch. 1993. A population needs-based approach to health care resource allocation and planning in Ontario: a link between policy goals and practice. C J Public Health 84:112-117.

McKelvey, W., J.G. Brody, A. Aschengrau, and C.H. Swartz. 2004. Association between residence on Cape Cod, Massachusetts, and breast cancer. Ann Epidemiol 14(2):89-94.

MDPH (Massachusetts Department of Public Health). 1999. Upper Cape Cod Cancer Incidence Review, 1986-1994 Final Report Appendices. Bureau of Environmental Health Assessment Environmental Epidemiology Unit. Boston: Boston University School of Public Health.

MDPH. 2001. Massachusetts Birth Defects, 1999. Bureau of Family and Community Health, Bureau of Health Statistics Research and Evaluation. Boston: Boston University School of Public Health.

MDPH. 2002. Technical Advisors' 3rd Report (2000-2001). Research on Health and Environmental Exposure Massachusetts Military Reservation, Louise Ryan, Joel Schwarz, Robert Goble, and Dale Hattis. Boston: Harvard School of Public Health.

MDPH. 2003. Cancer incidence and Mortality in Massachusetts 1996-2000: statewide report. Boston: Massachusetts Department of Public Health.

MITRE. 2000. MITRE technical report RF power density exposure at ground level for the PAVE PAWS radar at Cape Cod—questions and answers. MITRE Corporation, Bedford, MA.

Rothman, K.J. and S. Greenland. 1998. Modern Epidemiology. Philadelphia: Lippincott Williams and Wilkins.

US Census Bureau. Census 2000. http://www.census.gov/main/www/cen2000.html (November 2004).

Willett, W.C., B. Rockhill, S.E. Hankinson, D.J. Hunter, and G.A. Colditz. 2000. Epidemiology and nongenetic causes of breast cancer. Pp. 175-220 in Diseases of the Breast, Second Edition. J.R. Harris, M.E. Lippman, M. Morrow, and C.K. Osborne, eds. Philadelphia: Lippincott Williams & Wilkins.

ANNEX 9-1: DATA SOURCES

- Cancer incidence data from 1986 to 1994 for 24 cancer types in 30 census tracts in upper Cape Cod (*Upper Cape Cod Cancer Incidence Review 1986-1994[3]*)
- Peak and average PAVE PAWS power density from MITRE (*Map4&7*)
- Average PAVE PAWS power density from Broadcast Signal Lab—PAVE PAWS Average Exposure.RMX *[Map 8]*
- Digital elevation model (DEM) downloaded from the GeoComm International Corporation (http://gisdatadepot.com/dem *[Map2 & Map 3]*)
- Total population in 1990 and 2000 by gender and by age at block group level—Census bureau (web Table QT-P1A & P12) [*Map 5 & Table 1*]
- Total population in 2000 by census block in the form of GIS Datalayer from Massachusetts Geographic Information System (http://www.state.ma.us/mgis/cen2000_blocks.htm [*Map 9*])
- Poverty status in 1989 by gender and by age—Census bureau (web Table P118) [*Figure 2 & Figure 3*]

ANNEX 9-2: GEOBUGS CODE

Model association between cancer incidence and MITRE peak exposure controlling for SES

```
{
 # X : exposure
 # S : SES
 for (i in 1 : N) {
 O[i] ~ dpois(mu[i])
 log(mu[i]) ← log(E[i]) + alpha0 + alpha1 * X[i] + gamma*S[i]+ b[i]
 RR[i] ← exp(alpha0 + alpha1 * X[i] + gamma*S[i]+ b[i])
 SIR[i] ← 100*O[i]/E[i]
 SIR.smooth[i] ← 100*mu[i]/E[i]
 }
 # CAR prior distribution for random effects:
 b[1:N] ~ car.normal(adj[], weights[], num[], tau)
 for(k in 1:sumNumNeigh) {
 weights[k] ← 1
 }
 # Other priors:
 alpha0 ~ dflat()
 alpha1 ~ dnorm(0.0, 1.0E-5)
 gamma ~ dnorm(0.0, 1.0E-5)
 tau ~ dgamma(0.0001, 0.0001) # prior on precision
 sigma ← sqrt(1 / tau) # standard deviation}
```

10

Summary of Conclusions and Recommendations

RADAR CHARACTERISTICS

Both phased-array antennas and reflector antennas have time delays at wide angles. Arrays and dishes of comparable size have comparable delays. PAVE PAWS is a narrow-band system (5 MHz) with significant filtering for interference reduction. This results in waveform rise and decay times that are several times the maximum delay time of 60 nsec. Effects of delay time were not observed in the extensive Phase IV measurements.

The large number of PAVE PAWS active elements (1792), and their irregular spacing, make the discrete-beam formation almost indistinguishable from a continuous formation. There is no evidence to indicate that phased-array radar effects are different from those of dish radars; the Phase IV-measured waveforms affirm this.

Precursors are portions of the signal waveform that can travel faster or slower than the main part of the signal. Those precursors may occur in dispersive media for signals with large bandwidths. They have been calculated and measured, but only for wideband signals, typically, 10,000 MHz bandwidth, in dispersive media. They may decay slowly, but only after significant attenuation in the cellular media. With respect to narrow-band PAVE PAWS radar exposure, any precursors introduced into the human body would likely be below the noise level, and are unlikely to be of consequence. Precursor formation is directly related to bandwidth (rise time) in nsec, and dispersion, but not to slope (V/m/nsec).

Extremely-low-frequency (ELF) electric fields have been shown to produce biological effects at low-field intensities. Because PAVE PAWS is modulated at ELF frequencies, it is theoretically possible that exposure to PAVE PAWS radar could result in inducing very low levels of ELF fields in the body. To determine the possible levels of ELF energy induced by PAVE PAWS radar exposure, a Fourier analysis could be made of a representative search-pulse pattern, including the dead (maintenance) interval. This information, in combination with measurements of the non-linearities in human tissue impedance at the carrier frequency, would allow an estimation of the extent to which the sideband energy is converted to ELF energy in an exposed body and this in turn could be evaluated for possible biological effects if the ELF-exposure levels are large enough to correspond to observed effects at ELF frequencies. It should be noted that many radars employ diagnostic intervals in their waveforms that result in a small ELF modulation.

EXPOSURE LEVELS

Power-density measurements recorded by different groups at different times within the communities surrounding the Cape Cod PAVE PAWS radar are generally consistent with each other and with modeled results. They show distribution patterns that are strongly influenced by site-specific local topography and intervening terrain at any given location. The available data and models of the PAVE PAWS power-density emissions adequately characterize the spatial distribution of the exposures occurring throughout the communities of Cape Cod. The modeled power densities tend to be higher than measured power-densities, perhaps due to conservative assumptions made in the modeling effort.

BIOLOGY

No biological studies have been identified that used RF exposures identical to the PAVE PAWS system. However, a variety of biological studies have been undertaken based on short-term (hours), medium-term (weeks to months), and long-term (years) RF exposures in the frequency range of PAVE PAWS.

Short Exposure Times

Many short-exposure-time studies (hours to days) have addressed the possibility of DNA damage and other cellular responses following exposure to RF. DNA damage is a normal occurrence in living cells, with estimates that mammalian cells encounter thousands of DNA strand breaks/cell/day, most of which are repaired without adverse effects. In numerous short-term exposure studies, no reproducible effects on DNA damage have been observed, as measured by a number of different methods (micronucleus formation, DNA strand-break analysis, apoptosis, and others). While some studies have shown significant effects on gene expression due to modulated RF exposure of cells in culture, these do not include endpoints traditionally associated with carcinogenesis.

Intermediate Exposure Times

In studies involving exposure times on the order of weeks to months at levels of 900/1900 MHz, and in one case, 60 Hz, there are reports of statistically significant changes in development and differentiation of tissues and organisms. In particular, studies of exposures of chick embryos have shown a delay in development and reduced fertilization. A decreased developmental stability was also observed following exposure of fruit fly eggs. The implications of this observation for human embryonic development are unknown at this time.

Perhaps of greatest interest are the studies undertaken at 435 MHz (similar to the central frequency of the PAVE PAWS system), which demonstrated a substantial and significant reduction in dopamine levels in exposed animals.

Long Exposure Times

Long-term exposure studies (multiple years) are principally two-year bioassay experiments or carcinogenesis studies in animals and studies of changes in plant growth. A wide range of carcinogenesis studies have been conducted in animals exposed to RF fields. Such studies can generally be described as those with field intensities high enough to cause an increase in temperature in tissue (thermal) and those with field intensities below that level (athermal; comparable to the PAVE PAWS signal). Review of the lower-intensity animal studies identified some potential influences of exposure, but no reproducible indications of increased cancer risk with exposure have been shown.

Tree growth studies have been undertaken with radar exposures similar to those of PAVE PAWS at an early-warning radar installation in Latvia. Those studies showed a dose- and distance-dependent decrease in tree-ring width. Plant growth-response effects have been confirmed by changes in cell-wall synthesis activity after RF exposure.

Theoretical Mechanisms

At the present time there has been no physical mechanism shown to connect electromagnetic fields at the exposure levels associated with the PAVE PAWS radar to changes in chemical reactions that would lead to biological effects.

Biology Recommendations

Given the observation of long-term RF-exposure effects in plants and the several (though unreproduced) reports of effects in intermediate-duration exposures, the committee concludes certain additional studies on the biological effects of PAVE PAWS-specific RF exposures are warranted. In particular, studies using large-scale genomic screening for gene and protein expression at the cellular level and studies of plant growth in the area around the PAVE PAWS facility would improve our understanding of possible biological effects of long-term exposures to this radar.

While biological responses do not necessarily translate into health effects, it is important to follow-up on known biological effects to determine if they relate to an identifiable health effect. The committee concludes that there is a need for three types of studies:

1. Studies are recommended that include large-scale genomic and proteomic screening to identify gene- and protein-expression patterns in cell and animal studies after exposure to simulated PAVE PAWS radiation, both at levels approximating peak-exposure levels of the CAPE COD population and at higher power levels to identify potential threshold power densities for biological effects. The committee recognizes that the limitation of these studies is that they may have limited applicability to the human situation. However, the strength of these studies is that they will provide information covering a wide range of cellular activity in a limited number of experiments and the information gained may be useful in generating theoretical mechanism-related hypotheses of the possible effects of PAVE PAWS exposures.

2. Studies to further investigate the potential influence of PAVE PAWS exposure on neurotransmitter (e.g., dopamine) concentrations in the central nervous system are also recommended. The observation of a robust depression in dopamine levels (~50%) with an onset concurrent with the start of exposure, and lasting for the duration of the exposure, could prove to be of great importance if this effect can be replicated. Dopamine is closely involved in motor control (e.g., depressed levels of dopamine are believed to be causal in the etiology of Parkinson's disease) and so there could be a direct link between these observations and health effects. It is recommended that these studies be undertaken utilizing modulation frequencies more representative of the PAVE PAWS system rather than the 1 kHz utilized by Toler and others, as these lower modulation frequencies are closer to dominant brainwave frequencies.

3. Studies of tree growth in the vicinity of the PAVE PAWS facility are also recommended. Evaluation of tree-ring width, comparing rings from trees before and after the facility became operational with similar trees in areas that provide similar growth conditions outside the beam, should be possible with minimal or no impact on the environment. While these studies have the limitation of not being directly applicable to human health, they have the specific strength of involving long-term exposures (years). In addition, if the findings replicate the Latvian tree-growth results, it is anticipated that the information from these studies could lead to mechanism-generating hypotheses.

EPIDEMIOLOGY

Massachusetts Department of Public Health Statistics for Upper Cape Cod

Using the upper Cape Cod cancer incidence review of 1986-1994 from the Massachusetts Department of Public Health, statistically significant increases in the standardized incidence ratio (SIR) are seen for the following cancers: colorectal (SIR 112), breast (SIR 110), prostate (SIR 130), and lung (SIR 112) (Table 9-2 in Chapter 9).

As an overall measure of health for the upper Cape Cod towns, premature mortality before age 75 is a useful indicator. Based on 2001 data, Barnstable, Falmouth, Mashpee, and Sandwich have lower mortality than the Massachusetts state average, while Bourne has elevated mortality. The observed elevated cancer-incidence rates among residents of upper Cape Cod have not been adequately explained through subsequent investigations.

A comparison of the calculated SIR for 5 categories consisting of total cancers, breast, colon, lung, and prostate cancer during 1987-1994 versus 1995-1999 for 5 towns in upper Cape Cod (see Table 9-2) demonstrated no consistent pattern of increase during the latter time period. During these two time periods, a decrease was observed in 15 out of 25 SIRS, no change in 4 out of 25 SIRS, and an increase in 6 out of 25 SIRs, thus indicating that increasing duration of exposure to the PAVE PAWS radar has not resulted in increased incidence of cancer.

Human Epidemiology of Pulsed-RF Exposures

Studies of the U.S. Embassy in Moscow, the Radiation Laboratory at MIT, and of 40,000+ U.S. Navy personal all showed no health effects ascribable to the exposure to radars. Studies of the Polish military and the populations near Skrunda in Latvia can be considered as indicative of a possible need for follow-up studies. However, the difficulties in defining possible confounders reduces the weight given to these studies by the committee until they are supported by other epidemiological studies.

Committee Analysis of PAVE PAWS and Cancer Incidence

Peak and average power-density measurements obtained by use of (1) digital elevation maps, which take into account the height and operating angle of the radar, (2) MITRE data, and (3) the BSL report, all provided similar estimates of the exposure to the PAVE PAWS beam at each census tract within upper Cape Cod (see Figure 10-1 below) with the exception that the modeled estimates are consistently higher than the measured power densities.

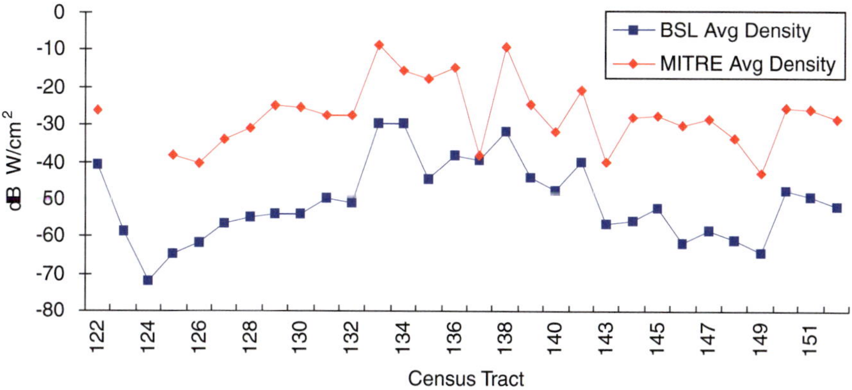

FIGURE 10-1 Comparison of the two sets of census tract specific average power-density estimates. Census tracts 123 and 124 have zero exposure estimates based on MITRE data and are omitted.

The analysis conducted by the NRC committee, based on 2000 data (Table 9-7 in Chapter 9), and using available information on population density, topography, and direction of the PAVE PAWS radar beam, estimated that the proportion of the population of upper Cape Cod residing (primary residence) in areas that result in direct exposure to the PAVE PAWS beam was 11.8% in 1990 and 12.4% in 2000.

The analysis conducted by the NRC committee of the occurrence of all cancers combined, as well as specific cancers, including colorectal, breast (female), prostate, and lung, did not identify any increase in cancer risk with estimated peak or average power-density exposure to the PAVE PAWS beam. Statistical adjustment for socioeconomic status, utilizing the percent of population below the poverty level, had little impact on the observed correlations (Figures 9-7 to 9-11 in Chapter 9).

Epidemiology Recommendations

The size of the population on upper Cape Cod is sufficient (with 10 years of observation) to design and conduct research, either prospective or retrospective, that could address increases in risk levels between 40% and 100% for some of the more common cancer outcomes (i.e., colo-rectal, lung, breast, prostate, bladder, and non-Hodgkin lymphoma; see Table 9-6 in Chapter 9).

Because of the limitations of exposure estimates, confounders, and range of health outcomes it is recommended that future investigations integrate (1) exposure assessment and health at the census-block level; (2) personal exposure characteristics other than just residential location; (3) extensive consideration of potential confounders; and (4) health outcomes other than cancer. In addition, it is recommended that geographical-correlation studies should be carried out for age-specific strata (e.g., under 30 or 45).

Appendix A

Interim Letter Report

THE NATIONAL ACADEMIES
Advisers to the Nation on Science, Engineering, and Medicine

National Academy of Sciences
National Academy of Engineering
Institute of Medicine
National Research Council

DIVISION ON EARTH AND LIFE STUDIES
Board on Radiation Effects Research

November 15, 2002

Lt. Col. Richard Ashworth, Chief
Bioenvironmental Engineering
Office of the Command Surgeon
Headquarters Air Force Space Command
150 Vandenberg Street, Suite 1105
Peterson AFB, CO 80914-4550

Dear Col. Ashworth,

The National Research Council has been asked to update its 1979 report on the analysis of exposures to and potential biologic effects of the PAVE PAWS radar system. This update will be completed at a later date after an extensive review of available data. The committee is to first determine whether available research data on continuous and pulsed radiofrequency (RF) energy are adequate for determining the biologic and potential health effects of the PAVE PAWS phased-array system (see Appendix, lines 1–3 of Statement of Task). The Air Force has asked that this determination be communicated as a letter report. The committee's initial assessment is that the data on continuous and pulsed RF energy are not adequate for determining the biologic and potential health effects of the PAVE PAWS phased-array system at this time. This conclusion is based on two observations.

First, issues related to the rise and fall portions of the PAVE PAWS waveform have been raised, but cannot be addressed, because we have available only preliminary, but inadequate, measurements of the PAVE PAWS waveform. These measurements are important to characterize transient aspects of PAVE PAWS radiation and to answer a question raised by others about the comparability of the PAVE PAWS phased-array radiation with radiation emitted by non-phased array systems. Hopefully, the significance of the PAVE PAWS wave shape will be clarified by the Phase IV measurement data .

Second, although more than 2 decades of exposure to PAVE PAWS phased-array radiation has been experienced by the Cape Cod population which should allow a direct evaluation of any adverse health effects, there are currently inadequate data about the distribution of population exposures in the Cape Cod region. Adequate exposure data would be necessary to carry out this direct evaluation of the effects of the PAVE PAWS waveform. The 1978 and 1986 PAVE PAWS power-density measurements are limited to 37 locations in proximity to the PAVE PAWS site. Apparently these measurements were not designed to cover a wide range of census tracts for epidemiologic purposes. Use of epidemiologic methods is considered by this committee to be important to the determination of possible health effects of the PAVE PAWS radar. Accordingly, adequate power-density measurements that discriminate the PAVE PAWS emissions and that cover relevant census tracts are deemed necessary to facilitate the determination of possible health effects. It is our understanding that power-density measurements will be commissioned by the PAVE PAWS Steering Group for use in their epidemiology effort and that this information may be available to our committee as well. The committee considers access to these data to be important to items 2 and 3 of the committee's Statement of Task (see Appendix, items 2 and 3 of the Statement of Task). After the phase IV data are evaluated and during the committee's update of the 1979 National Research Council report, the committee will re-evaluate the adequacy of the existing RF energy literature for determining the potential biologic and health effects of the PAVE PAWS phased-array system. The committee further expects that the Phase IV measurement data, additional power density measurements, and computed exposure estimates that include a propagation model will be extremely useful in its attempts to evaluate the potential health effects of the PAVE PAWS radar on the Cape Cod population. The committee recognizes that much research has been conducted over the past 23 years that may provide insight into the possible biologic and health effects of PAVE PAWS radiation, and the committee will be reviewing the cellular, animal, and epidemiologic RF exposure and health effect data in detail. To this end the committee has constructed a set of potentially relevant RF field characteristics that will be used as guidelines in the identification of experiments in the literature that could be relevant to the update of the 1979 Research Council report (see Table).

The ranges described are to indicate the focus of the committee's review, though the committee may determine that certain data obtained under other exposure conditions are relevant to its task. It is beyond the scope of this letter report

Characterization of the PAVE PAWS Waveform and Identification of Waveform Characteristics Suitable for the Evaluation of Biologic and Potential Health Effects of PAVE PAWS Radar

Characteristic	PAVE PAWS	Relevant Range of Data to Be Considered by the Committee	Comments
Frequency	420-450 MHz	10 MHz-2 GHz[a]	There are common mechanisms of interaction in the 10 MHz-2 GHz frequency range
Incident Average Power Density	<280 μW/cm^2[b]	<1 mW/cm^2 or SAR <1 mW/g (<1 W/kg)	Specific Absorption Rate increases for a given incident power with increasing frequency
Modulation	Pulse	Pulse (CDMA, TDMA, IDEN, GSM, etc.)	To include all existing RF pulse-modulation technologies
Pulse Width	0.25-16 ms	0.001-100 ms	
Pulse Repetition Rate	0.02-20 Hz	0.01 Hz-10 kHz	
Pulse Rise Time (τR)	>20 ns	10 ns < τR < 10 μs	τR should be less than the charge relaxation time (ε/σ for biologic tissue, where ε is the dielectric permittivity of tissue, and σ the conductivity. Max estimated relaxation time is 10 ms.
Exposure Duration	Continuous	Continuous-1 hr/day	

[a]Upper limit is intended to include cell-phone-frequency research (which goes as high as 1.9 GHz). The 2.45 GHz research in the literature is continuous RF energy and would be excluded by the modulation criteria (pulsed exceptions will be considered).

[b]The MITRE report on PAVE PAWS (2000) indicates that at the perimeter fence line the energy level does not exceed the IEEE thirty-minute average limit for human exposure. The IEEE limit is 280 microwatts per square centimeter.

to list all possible sources of data that fall within the committee's proposed criteria to be reviewed. The committee is using a database with more than 37,000 references in addition to the information that various members have acquired in working in the field of bioelectromagnetics for many years. It should be further noted that the extensive literature on power-frequency fields is not considered relevant to the evaluation of PAVE PAWS health effects in our study because the physical mechanisms of interaction between power-frequency fields and tissue, on the one hand, and RF energy and tissue at low field strengths, on the other, are expected to be different. Furthermore, studies of power-frequency fields tend to concentrate on the effects of magnetic fields whereas studies on the effects of RF energy focus on power-density measurements or SAR's. In addition, the committee proposes to focus on pulsed modulated fields rather than continuous wave fields. Continuous wave RF exposure experiments do not involve the same broad spectrum of energy and also utilize far lower peak exposure levels to attain the same average exposure levels. As a result, inclusion of continuous wave studies in the literature review may unfairly bias the observed results against any potential effects of exposure.

References Cited

MITRE. 2000. Kramer, A.G., B.P. Nelson, and R.E. Wakefield. RF Power Density Exposure at Ground Level for the PAVE PAWS Radar at Cape Cod - Questions and Answers. Bedford, MA.

NRC (National Research Council). 2002. Letter Report to the Department of the Air Force from the Committee to Assess Potential Health Effects from Exposure to PAVE PAWS Low-level Phased-array Radiofrequency Energy: Recommendations for Phase IV Measurements. Washington, D.C.: National Academy Press http://www.nap.edu/books/NI000468/html/

Sincerely,

Frank Barnes, Ph.D.
Chairman
Committee to Assess Potential Health Effects from Exposures to
PAVE PAWS Low-level Phased-array Radiofrequency Energy

COMMITTEE TO ASSESS POTENTIAL HEALTH EFFECTS FROM EXPOSURES TO PAVE PAWS LOW-LEVEL PHASED ARRAY RADIOFREQUENCY ENERGY

FRANK E. BARNES (Chair), University of Colorado at Boulder, Boulder, CO
ROBERT C. HANSEN (Vice Chair), R. C. Hansen, Inc., Tarzana, CA
LARRY E. ANDERSON, Battelle Pacific Northwest National Laboratories, Richland, WA
GRAHAM A. COLDITZ, Harvard Medical School, Boston, MA
FRANCESCA DOMINICI, Johns Hopkins University, Baltimore, MD
KENNETH J. McLEOD, State University of New York, Binghamton, NY
KEITH D. PAULSEN, Dartmouth College, Hanover, NH
LESLIE L. ROBISON, University of Minnesota Cancer Center, Minneapolis, MN
SUSAN L. SANTOS, University of Medicine and Dentistry of New Jersey and East Orange, War Related Illness and InjuryStudy Center, Medford, MA
JAN A. J. STOLWIJK, Professor Emeritus, Yale University, New Haven, CT
GAYLE E. WOLOSCHAK, Northwestern University, Chicago, IL

NATIONAL RESEARCH COUNCIL STAFF

EVAN DOUPLE, Study Director
RICK JOSTES, Study Director
TAJUANA CLAYTON, Project Assistant
DORIS E. TAYLOR, Staff Assistant

SPONSOR'S PROJECT OFFICER

LT. COL. RICHARD ASHWORTH, U.S. Air Force

APPENDIX

An Assessment of Potential Health Effects from Exposures to PAVE PAWS Low-level Phased-array Radiofrequency Energy

Statement of Task

The committee will first determine whether continuous and pulsed radiofrequency (RF) energy research data are adequate for determining the biological and potential health effects of the PAVE PAWS phased-array system. This determination will be communicated to the sponsor as a letter report. If the research data from continuous and pulsed RF energies are considered to be applicable for the determination of possible health effects of phased-array RF energy, the committee will use this information to update the 1979 National Research Council analysis of the exposure levels and potential biologic effects of the PAVE PAWS Radar System. If the data are not applicable, the committee will use other information that it determines to be relevant to phased array health effects to update the 1979 report. In this update the committee will evaluate potential biological and health effects, evaluate exposures of the public to electromagnetic energy from the PAVE PAWS system, and make recommendations for the need for, and focus of, additional scientific studies to address continued scientific uncertainty related to health outcomes and exposure to low-level radiofrequency energy emitted by the PAVE PAWS Radar System. At the completion of its work, the committee will provide an update of the 1979 Research Council report that includes a discussion of:

1. The applicability of, and the level of uncertainty associated with, using data derived from cell, animal, and epidemiological studies employing continuous-wave exposure for evaluation of potential adverse health effects following phased-array exposures;
2. The extent of the exposure of the public to electromagnetic energy from the PAVE PAWS system;
3. Potential biological and health effects of the PAVE PAWS Radar System, and
4. Recommendations for appropriate follow-on study design issues, including the strengths and limitations of the approaches suggested and the potential value of the proposed work.

REFERENCE

NRC (National Research Council). 1979. Analysis of the Exposure Levels and Potential Biologic Effects of the PAVE PAWS Radar System. Washington, DC: National Academy Press.

Appendix B

Acronyms and Abbreviations

α	polarizability; or lowest frequency dispersion
AFB	Air Force Base
AFRL	Air Force Research Laboratories (at Brooks Air Force Base, San Antonio, Texas)
AGNIR	Advisory Group on Non-ionizing Radiation
ANSI	American National Standards Institute
B	magnetic flux density
B-dot	sensor that measures the derivative of a magnetic field
BBB	blood-brain barrier
BP	blood bressure
BRFSS	Massachusetts Behavioral Risk Factor Surveillance System
BSL	Broadcast Signal Laboratories
β	second frequency dispersion
CA	Canada
Ca^{++}	calcium ion
CCCR	Cape Cod Commission Reporter
CCEC	Cape Cod Environmental Coalition
CDMA	Code Division Multiple Access (cellular telephone technology originally known as IS-95)
CI	confidence interval
CNS	central nervous system
COPPS	Coalition for the Operation of PAVE PAWS Safely
CRC	CRC Press, Boca Raton, Florida
CW	continuous wave

D	reflector diameter $\left[\text{in } \Delta t = \frac{D/\lambda}{2F}\frac{(\cos\theta}{8f/D} + \sin\theta)\right]$
D-dot	sensor that measures the derivative of an electric field
Da	dalton
dB	decibel
DC	direct current
DDT	dichloro-diphenyl-trichloroethane
DED	Directed Energy Directorate (of the Air Force Research Laboratories at Brooks Air Force Base, San Antonio, Texas)
DEM	Digital Elevation Model
DMBA	7,12-dimethylbenz[a]anthracene
DNA	deoxyribonucleic acid
dsb	double-strand breaks
°	degrees
Δf	bandwidth (delta frequency)
E	electric field
EEG	electroencephalogram
EHS	Environmental Health and Safety
EIG	Engineering Installation Group (Keesler Air Force Base, Biloxi, Mississippi)
EIS	Environmental Impact Statement
ELF	extremely low frequency
EMF	electro-magnetic field
EPA	Environmental Protection Agency
f	frequency; focal length $\left[\text{in } \Delta t = \frac{D/\lambda}{2F}\frac{(\cos\theta}{8f/D} + \sin\theta)\right]$
f_{rs}	principal frequency of the rise time
F	force; frequency $\left[\text{in } \Delta t = \frac{D/\lambda}{2F}\frac{(\cos\theta}{8f/D} + \sin\theta)\right]$
FDMA	Frequency Division Multiple Access (cellular telephone technology)
FM	frequency modulation
ft	feet
γ	third frequency dispersion
GAO	General Accounting Office
Gy	gray
cGy	centigray (10^{-2} gray)
GPS	Global Positioning System
GSM	Groupe Spécial Mobile (cellular telephone technology)
h	Planck's constant

HCN	Health Council of the Netherlands
HED	Human Effects Division (of the Air Force Research Laboratories at Brooks Air Force Base, San Antonio, Texas)
HPLC	high performance liquid chromatography
HR	heart rate
hsp70	heat-shock protein 70
Hz	hertz
kHz	kilohertz (10^3 hertz)
MHz	megahertz (10^6 hertz)
GHz	gigahertz (10^9 hertz)
THz	terahertz (10^{12} hertz)
ICNIRP	International Commission on Non-Ionizing Radiation Protection
ILHI	Incomplete Lipschitz-Hankel Integrals
IR	infrared
IEEE	Institute of Electrical and Electronic Engineers
IITRI	Illinois Institute of Technology Research Institute
J	drift current density; or joule (kJ = kilojoule = 10^3 joule)
k	Boltzmann's constant
kg	kilogram (10^{-3} gram)
λ	wavelength $\left[\text{in } \Delta t = \frac{D/\lambda}{2F}\frac{(\cos\theta}{8f/D} + \sin\theta)\right]$
m	meter
cm	centimeter (10^{-2} meter)
km	kilometer (10^{-3} meter)
MA	Massachusetts
Mag	magnitude
MDPH	Massachusetts Department of Public Health
MISER	Massachusetts Institute for Social and Economic Research
MMR	Massachusetts Military Reservation
MPE	Maximum Permissible Exposure
MS	Mississippi
μ	mobility [in $J = N\mu\alpha V(E\bullet\nabla)E$]
N	particle density [in $J = N\mu\alpha V(E\bullet\nabla)E$]
NA	National Academies
NAS	National Academy of Sciences
Nc	not calculated

NCRP	National Council on Radiation Protection and Measurements
NIEHS	National Institute of Environmental Health Sciences
NRC	National Research Council
NRPB	National Radiological Protection Board
NSWC	Naval Surface Weapons Center
NYU	New York University
PAVE PAWS	"PAVE" is an Air Force program name and "PAWS" stands for Phased Array Warning System
PCE	perchloroethylene
PI	Physics International
PPPHSG	PAVE PAWS Public Health Steering Group
%	percent
Φ	azimuth angle
q	charge
Q	ratio of stored energy to dissipated energy
QPSK	Quadrature Phase Shift Keying (a digital frequency modulation technique for sending data over coaxial cable or radio)
R	resistance [in $\langle V^2 \rangle = 4kT\Delta fR$]; or distance [in $1/R^2$]
RF	radiofrequency
RFEMF	radiofrequency electromagnetic field
RG-8	type of coaxial cable
RNA	ribonucleic acid
S	siemen (conductivity is defined in S/m)
SAR	Specific Absorption Rate
sec	second
msec	millisecond (10^{-3} second)
nsec	nanosecond (10^{9} second)
psec	picosecond (10^{-12} second)
SES	socio-economic status
SIR	standard incidence rates or standardized incidence ratio
SMR	standard mortality rates or standardized mortality ratio
SRI	Stanford Research Institute (type of dish antenna)
ssb	single-strand breaks
SSI	Swedish Radiation Protection Institute
σ	conductivity (is defined in S/m)
T	tesla; or absolute temperature [in $\langle V^2 \rangle = 4kT\Delta fR$]
mT	millitesla (10^{-3} tesla)
TEM	Transverse Electro-Magnetic (type of antenna)

TPA	12-O-tetradecanoylphorbal-13-acetate
TR	Transmit-Receive modules
TV	television
θ	polar angle
τr	pulse rise time
UHF	ultra-high frequency
UK	United Kingdom
US	United States
UV	ulraviolet
UVA	type of ultraviolet light
UVB	type of ultraviolet light
v	velocity
V	volt; or volume of the ion or molecule in [in $J = N\mu\alpha V(E\bullet \nabla)E$]
kV	kilovolt (10^3 volt)
VHF	very-high frequency
W	watts; or energy [in W = hf]
kW	kilowatts (10^3 watts)
W/cm^2	watts per centimeter squared (power density)
$\mu W/cm^2$	microwatts (10^{-6} watts) per centimeter squared (power density)

Committee Biosketches

FRANK S. BARNES, Ph.D. (*Chair*), is Distinguished Professor in the Department of Electrical and Computer Engineering at the University of Colorado at Boulder. He received a Ph.D. in Electrical Engineering from Stanford University. His career has included research in a wide variety of applications in physics and electrical engineering, focusing on fundamental research on the biological effects of electromagnetic fields, surgical procedures, and telecommunications education. His research has included the effects of radio waves, fields from power lines, and ultrasonic fields on biological systems—trying to understand the mechanisms of interaction that might lead to identification of hazards, the setting of safety standards. Dr. Barnes is an AAAS Fellow, received the Centennial Award and Third Millennium Medal of the Institute of Electrical and Electronics Engineers, and was elected to the National Academy of Engineering in 2001. He currently is an Ex Officio Member of the U.S. National Committee for the International Union of Radio Science and a past-president of the Bioelectromagnetics Society. In 2004 he was awarded the Bernard M. Gordon Prize by the National Academy of Engineering.

ROBERT C. HANSEN, Ph.D. (*Vice-Chair*), received his doctorate in electrical engineering from the University of Illinois. Since 1971, he has been a consulting engineer for electromagnetic antennas and systems-related problems and is recognized as an expert in near-field phased-array radar systems. In 1960, Dr. Hansen became a senior staff member in the Telecommunications Laboratory of STL (now TRW), and was engaged in communication satellite telemetry, tracking, and command. From 1964 to 1966 he formed, and was director of, the Test Mission Analysis Office responsible for computer programs for the planning and control of classified Air Force satellites. He is a Fellow of IEEE and IEE, a registered professional engineer in California and England, and a member of the American Physical Society and Sigma Xi, the Scientific Research Society, and Tau Beta Pi, the Engineering Honor Society. He was awarded an honorary Doctor of Engineering degree by the University of Missouri-Rolla in 1975 and was elected to the National Academy of Engineering in 1992. Dr. Hansen has written over 100 papers on electromagnetics and is the editor of several books, including the 3-volume *Microwave Scanning Antennas*, and author of *Phased Array Antennas*. He received the IEEE 2002 Electromagnetics Award.

LARRY E. ANDERSON, Ph.D., received his Ph.D. in Biochemistry (neurochemistry) from the University of Illinois. Dr. Anderson was program manager for research efforts investigating the interactions between biological systems and non-ionizing radiation at Battelle, Pacific Northwest National Laboratory and has a faculty position at Washington State University in Biological Science. He is a specialist in neurochemistry whose research has included investigations on neurohormones and associated circadian biochemistry as well as the carcinogenecity of environmental physical agents principally power-frequency fields, radiofrequency energy, and visible light. He is a past-president of the Bioelectromagnetics Society and has served on a number of national and international committees on non-ionizing radiation and health, including the National Council on Radiation Protection; International Conference on High Voltage Systems; International Agency for Research on Cancer Working Group on Extremely Low Frequency Electromagnetic Fields; International Union of Radio Science, Commission K; International Commission on Non-ionizing Radiation Protection; National Institute of Environmental Health Sciences, Working Group on Biological Effects of Static and ELF Electric and Magnetic Fields; and the World Health Organization Working Group on Health Protection, Non-ionizing Radiation.

GRAHAM A. COLDITZ, M.D., Dr.P.H., received his M.D. from the University of Queensland, Australia, and his Dr.P.H. from Harvard. He is currently a Professor of Medicine, Harvard Medical School, Professor of Epidemiology, Harvard School of Public Health, and Epidemiologist, Brigham and Women's hospital. Dr. Colditz has a major interest in the etiology and prevention of cancer and is working with the Massachusetts Department of Public Health to translate research findings from ongoing cohort studies into public health strategies for prevention. He is the author or co-author of 461 original reports and 85 reviews. One of his publications is among the top 5 most-cited breast cancer papers of the 1990s.

FRANCESCA DOMINCI, Ph.D., received her Ph.D. in Statistics from the University of Padua, Italy. She is currently an Associate Professor in the Department of Biostatistics, Bloomberg School of Public Health, Johns Hopkins University.

Dr. Dominici's research has concentrated on time-series analysis in environmental epidemiology, models for meta-analyses, and imputation techniques for incomplete data sets. Dr. Dominici is a member of the American Statistical Association and International Biometric Society. Dr. Dominici is the recipient of the Walter A. Rosenblith New Investigator Award, the Health Institute, Boston, MA, and the 2001 Young Investigator Award of the Statistics in Epidemiology Section of the American Statistical Association.

KENNETH J. McLEOD, Ph.D., obtained his doctorate in Electrical Engineering from the Massachusetts Institute of Technology and did his postdoctoral work in Cell Biology at Tufts University School of Medicine. Dr. McLeod is presently Professor and Chairman of the Department of Bioengineering at Binghamton University. Dr. McLeod's primary research interests are directed toward understanding the mechanism of interaction of biophysical factors in the processes of tissue development, healing, and adaptation, including endogenous and exogenously induced electromagnetic fields. Dr. McLeod is a past-president of the Bioelectromagnetics Society and the Society for Physical Regulation in Biology and Medicine. He received the Kappa Delta Award for Outstanding Bioelectrical Research in 1990 and the State University of New York Award for Excellence in Invention, Creation, and Discovery in 2004.

KEITH D. PAULSEN, Ph.D., obtained his doctorate in Biomedical Engineering from Dartmouth College and was an Assistant Professor in the Department of Electrical and Computer Engineering at the University of Arizona from 1986 to 1988. He is currently Professor, Thayer School of Engineering, and Director, Radiobiology and Bioengineering Research Program, Norris Cotton Cancer Center, Dartmouth Hitchcock Medical Center. He is a recognized expert in modeling and calculating the deposition of radiofrequency energies in various media and in tissues. Dr. Paulsen received the Morgan Parker Memorial Fellowship for potential contribution in biomedical engineering. He has served on technical review groups at the National Cancer Institute, the National Science Foundation, the U.S. Army Breast Cancer Research Program, and the National Institutes of Health. He is currently a member of the Radiation Study Section, National Institutes of Health.

LESLIE L. ROBISON, Ph.D., received his Ph.D. in Public Health from the University of Minnesota. Dr. Robison is Director, Division of Pediatric Epidemiology/Clinical Research; Associate Director, University of Minnesota Comprehensive Cancer Center; and Professor, Department of Pediatrics, School of Medicine and Division of Epidemiology, School of Public Health. Dr. Robison's research has focused on the epidemiology of pediatric malignancies and long-term outcomes of pediatric cancer survivors. He is Associate Chair of the Children's Oncology Group and principal investigator of the Childhood Cancer

Survivor Study. He holds the Children's Cancer Research Fund Endowed Chair in Pediatric Cancer, University of Minnesota Cancer Center (1993–present). Dr. Robison is an author of over 200 publications and has served on numerous boards, committees, and the NIH Epidemiology and Disease Control Study Section.

SUSAN L. SANTOS, Ph.D., M.S., has an M.S. in Civil Engineering and Public Health from Tufts University and a Ph.D. in Law, Policy, and Society (Concentration in Risk Communication) from Northeastern University. Dr. Santos is currently Assistant Professor in the Division of Health Education and Behavioral Sciences at the University of Medicine and Dentistry of New Jersey's School of Public Health. She is also Director, Risk Communication, for the War Related Illness and Injury Study Center in the East Orange VA Medical Center. She is owner and president of FOCUS GROUP Risk Communication and Environmental Management, a consulting firm, and is a Senior Faculty Member at Boston College Carroll School of Management. In the past, she served as Research Program Director at the Columbia University Center for Risk Communication, Director of Risk Assessment Services for ABB Environmental, and was an environmental engineer at the Environmental Protection Agency (EPA) where she won the EPA Bronze medal (1981) and the EPA Sustained Superior Performance Award (1984). Dr. Santos is a member of the Society for Risk Analysis, the Massachusetts Public Health Association, The American Public Health Association (sections on Environment and Epidemiology), and the International Association of Public Participation Professionals.

JAN A.J. STOLWIJK, Ph.D., is the Susan D. Bliss Professor of Epidemiology and Public Health, Emeritus, at the Yale University School of Medicine. He was chair of the Department of Epidemiology and Public Health from 1981 to 1989 and from 1993 to 1994. Dr. Stolwijk has been a member of a number of National Research Council committees studying possible health effects of electromagnetic fields including the Committee to Assess Possible Health Effects of Ground Wave Emergency Network, the Committee to Assess Possible Health Effects of Exposure to Residential Electric and Magnetic Fields, and the Committee to Evaluate Research on Power-Frequency Fields Completed Under the Energy Policy Act of 1992. He has also served as a member of the EPA Science Advisory Board and committees dealing with indoor air pollution and with particulate air pollution.